GEOMETRY IN ARCHITECTURE

GEOMETRY IN ARCHITECTURE

WILLIAM BLACKWELL, A.I.A.

KEY CURRICULUM PRESS
Innovators in Mathematics Education
P.O. Box 2304 • Berkeley • California • 94702

Library of Congress Cataloging in Publication Data:

Blackwell, William, 1929-
    Geometry in architecture

    Includes index.
        1. Architecture—Composition, proportion, etc.
2. Geometry.   I. Title.
NA2760.B53   1983   729   83-10281
ISBN 1-55953-018-9

Designed by Kenny Beck
Cover photo: Stairway in the Cafe Aubette, Strassbourg, by
        Jean Arp and Sophie Taeuber-Arp, 1928.
            (Photo courtesy Foundation Arp, Clamart, France.)

To Marjorie

PREFACE

ARCHITECTURE AND GEOMETRY are inseparable. Understanding the properties and relationships of lines, patterns, surfaces, and solids is vital to an architect, whether one is designing a coffee table or a fifty-story building, the Taj Mahal or a power plant. At the same time, without architecture, geometry would be a "dead language," a scholarly subject confined to textbooks and classrooms.

Together, architecture and geometry form a perfect union of creativity and discipline, the one the instrument for the other, a balance of imagination and exacting realism. Functioning together at their "highest and best," the art and the science produce structures of superb beauty and rich design that bring qualities of order, symmetry, and harmony to our lives.

Since geometry is encountered in the first written records of mankind and was developed extensively by the ancient Babylonians, Egyptians, and Greeks, one might suppose it had been exhausted of discoveries long ago. Not so! The subject is slow to reveal its secrets. There are still geometric principles to be uncovered and new applications that can improve our man-made environment. There are also principles that have been long forgotten or ignored that have strong relevance to architecture today.

Surprisingly, geometry courses are not a standard part of the student architect's curriculum. Architectural schools assume students already know the subject or will learn on their own the principles they need to know. I believe that one day geometry will be taught to architectural students. The subject is too relevant, useful, and exciting to be ignored.

This book represents a lifelong fascination with the endless relationships of geometric forms and the continual discovery and rediscovery of their applications to architecture and design. It formulates much that is intuitive. Although it was not intended initially, the study of geometry in architecture leads inevitably to a renewed appreciation of nature's own special symmetries and forms.

Each chapter begins with a summary of familiar principles, followed by new observations and insights which I believe will interest architects, designers, planners, engineers, mathematicians, and everyone who enjoys the geometric world. Among the new observations and insights are: the two-sided polygon, the progression of fourths, the relationship of wall height to floor area for optimum enclosure, the diamond and diagonal geometry of city planning, and the 3–4–5 and 6 idea.

Although this is not primarily a textbook, it is arranged in a logical progression, beginning with the plane shapes: right triangle, regular polygon, circle, and the most widely used shape in all of architecture, the ubiquitous rectangle. These, in turn, are followed by the solid shapes: prism, pyramid, sphere, and classical solid. All are presented in the context of architectural design.

The mathematics in the book, for the most part, are simple. It is rational simplicity, not complexity, which often makes things work, never more so than in architecture and city planning.

William Blackwell
Piedmont, California
January 1991

ACKNOWLEDGMENTS

THIS BOOK began many years ago in a classroom of the Anchorage, Kentucky public schools. The course was geometry, and the teacher was Edith Wood. She was a dedicated teacher who knew geometry as a part of life and knew the value of the basic geometric "tools" she taught so well. She let us, her students, experience the satisfaction that comes from developing complete solutions to difficult problems, and in that she established a lifelong pattern.

I gratefully acknowledge, as well, the teaching and influence of M.I.T. Professors Gyorgy Kepes, Richard Buckminster Fuller, and Lawrence Anderson, which readers will recognize in many places. Later in life, Barclay Jones, Murray Persky, Polly Roberts, and Ken Bath all contributed to the making of this book. I am especially indebted to Ken for making available several papers from his unique private collection which I would not have found otherwise.

The illustrations in the book have come from many sources: museums, galleries, foundations, corporations, publishers, photographers, and individuals. My thanks go to all of those who provided the prints and the permission to use them. The designs, models, and drawings not otherwise identified in the captions are my own. Photographers Jerry Ratto and Colin McRae considerably enhanced the quality of the geometric models.

My friends and colleagues James Ream and Richard Broder took the time to review the first draft and provide constructive comment. Wm. Dudley Hunt, Jr., architecture editor for John Wiley & Sons, added sound advice, linked with patience and graciousness.

Finally, I have often wondered why so many men dedicate books to their wives. Only now do I appreciate the extent of their participation and sacrifice. I am indebted.

W . B .

CONTENTS

GEOMETRY IN ARCHITECTURE

C H A P T E R 1

INTRODUCTION

Well then, tell me now, in anticipation, what characterizes a real architect?

First of all a poetic imagination; second, a broad sympathy, humane character, common sense and a thoroughly disciplined mind; third, a perfected technique; and, finally, an abundant and gracious gift of expression.

Then you don't value logic.

It has its excellent uses.

Louis Sullivan
Kindergarten Chats
1918

Figure 1.1
Relativity, a lithograph by M. C. Escher, 1953.
(Copyright © Beeldrecht, Amsterdam/VAGA,
New York. Collection Haags
Gemeentemuseum—The Hague, 1981; image
courtesy the Vorpal Galleries, New York, San
Francisco, and Laguna Beach.)

GEOMETRY, according to the *Encyclopedia Britannica*, is the study of space, and architecture, in the broadest sense of the word, is the creation of space by construction or subdivision. The two disciplines are virtually inseparable with one distinction. Geometry can exist without architecture but architecture cannot exist without geometry.

The simplest space enclosure, an igloo or tepee, involves a sure sense of elementary geometry—otherwise the parts will not fit together. Geometry is not all of architecture but it is an essential part of it.

The geometry that is the subject of this book is that of existent space, the geometry with which we are most familiar, the three-dimensional space of constructed objects. It is termed Euclidean after its most famous expositor, and relies to a great extent on the Cartesian coordinate system of right angle lines and planes. The geometry of this book is the framework within which architects must build.

ARCHITECTURE, PAINTING, AND SCULPTURE

Although closely related, the geometry of architecture is different from the geometry of painting and sculpture. Architectural space must serve the needs of humans with some exactness: floors must be level, stairs must be straight, the laws of gravity must be respected, buildings must be buildable. Hence the marvelous fantasy geometry of M. C. Escher (Figure 1.1) transcends the world of the practicing architect.

In a similar way, the highly geometric, colorful, and volatile paintings of Victor Vasarely (Figure 1.2) violate an essential requirement of architectural geometry, that architectural surfaces be built up through the repetition of identical pieces and processes, such as the laying up of a brick wall. If every piece is different, both the manufacturing process and the assembly process are impractical. These processes are normally learned and perfected for use on more than just one building.

In the recent past, art and architecture did come together in geometry. The paintings of Mondrian and the architecture of Mies van der Rohe both relied on refined proportions of rectangles and squares. To an architect, of course, the rectangle is

Figure 1.2
Vega, a painting by Victor Vasarely, 1957. (Photo courtesy of the artist, copyright S.P.A.D.E.M., Paris.)

a basic tool, but it is rare that an artist with the fluidity of paint at his command remains enchanted with the right-angle discipline as did Mondrian.

NATURE AND ARCHITECTURE

The enjoyment and appreciation of nature as an experience leads, naturally enough, to associations with architecture. These associations may be romantic but there is, at least, the wish that the two be one and inseparable, that constructed works be perfectly in tune with nature.

Fulfillment of this wish is made difficult by the fact that the purposes and processes of building are not the same as the purposes and processes of nature. The biological processes of nature obviously are different from building processes, and there are other important distinctions.

The relationship of building to nature is first of all defensive. A building must be designed to resist the onslaughts of nature—rain, wind, fire, snow, earthquake, temperature extremes, and so on. An architect who fails here will face a disgruntled client and probably a lawsuit. In the case of fortresses and, unfortunately, many other kinds of buildings, the architect must design for human onslaughts as well as those of nature. These primary purposes of shelter and protection dictate specific materials and geometries depending on the resources available. Roofs are sloped or domed, walls are braced at right angles, openings are framed or arched, and so on, not in contradiction to nature but with the intention of providing a safer and more comfortable environment than that found in nature.

A building is not a living thing although it is sometimes said to have "life" of a kind. It does no work and needs no muscle but it breathes through its ventilation system, has power in its veins, a body temperature, a circulation system, possibly a heart, and a skeleton covered and protected by a skin. It is, however, inanimate. Its "life" is the life of its occupants and it is that life which provides its reason for existence.

All living things adapt to their surroundings and, although buildings are inanimate, it is on this point that nature and architecture may have the most in common. In their designs most architects take advantage of terrain, views, prevailing winds, surface drainage, available sunlight and shade, color, and other aspects of the surroundings. Buildings may open outwardly as well as inwardly. They may encompass the landscape as well as provide shelter. Buildings can blend into the surroundings with a minimum of disruption; they can conserve natural resources. An example of the completely harmonious integration of rectangular shapes into a natural setting is Falling Water, a dwelling designed by Frank Lloyd Wright in 1936 (Figure 1.3). Rarely is such a natural site available. The major task today is to find the harmonies of the urban environment.

Whatever else, a building is not amorphous, unstructured, or accidental, characteristics sometimes found in nature. At its best, the design process is conscious, deliberate, rational and responsible. The geometry of architecture then suits the purposes of building.

Figure 1.3
Falling Water, a dwelling in Bear Run, Pennsylvania, by Frank Lloyd Wright, 1936. (Photo by Hedrich-Blessing.)

DISTINCTION BETWEEN PLANE AND SOLID GEOMETRY

In architecture, the rules of both plane and solid geometry apply to three-dimensional shapes and spaces. When height or thickness is added, plane geometrics on paper become rooms, buildings, and open spaces. Even flat, two-dimensional tile patterns have the unseen dimension of thickness when used in building. The tiles themselves are prismatic solids and must be designed and detailed as such. A multistory building consists of a number of two-dimensional floor plans which become solid shapes when assembled.

Strictly speaking, there is little that can be called plane geometry in architecture, although the separation between plane and solid geometric shapes, as followed (with some exceptions) in this book, continues to be useful.

The term "solid geometry" also can be misleading in architecture, since three-dimensional rooms, buildings, and open structures are not solids but voids suitable for human habitation and use. The volumes of architecture follow the rules of solid geometry, however, and for that reason the term is used.

The emphasis in architecture is usually on the creation and definition of *space* for rooms, galleries, city squares, and other purposes. Occasionally the emphasis is on the *molding* of a shape for a skyscraper or other purpose, in which case the architect's work is similar to that of a sculptor. In both processes, plane and solid geometries are fully utilized and frequently intermixed.

EXAMPLES OF GEOMETRY IN DESIGN

The constructed environment is the province of designers, planners, engineers, and architects. At the least, it encompasses city and regional planning, the design and engineering of buildings, and the design of furnishings. Geometry is a vital part of these activities. The following few examples illustrate the point.

Geometry in Furniture Design

An example of geometry in furniture design is the chair (Figure 1.4) designed by Gerrit Rietveld in Holland in 1917/18. The design emphasized the

Figure 1.4
Chair by Gerrit Rietveld, 1917/18. (Crown copyright Victoria and Albert Museum, London.)

rectangle and its extension into space. Contemporary designers have long since softened, and made more comfortable, designs for chairs but the influence of Rietveld's design was extensive. The chair has a red back, blue seat, and black stained members with painted yellow ends, colors which enhanced the proportions of the squares and rectangles as Mondrian was later to do in his geometric paintings.

Geometry in Structural Design

Figure 1.5 shows a long beam evenly weighted from end to end. If the supports are at the ends, the beam will sag in the middle, but if the supports are moved a certain distance from the ends, the weight will be balanced over the supports and sag will be greatly reduced. In this case, balance in design leads to economy and energy conservation.

Figure 1.6 shows a more elaborate structural pavilion with supports geometrically arranged to minimize the effect of the vertical loads. The long,

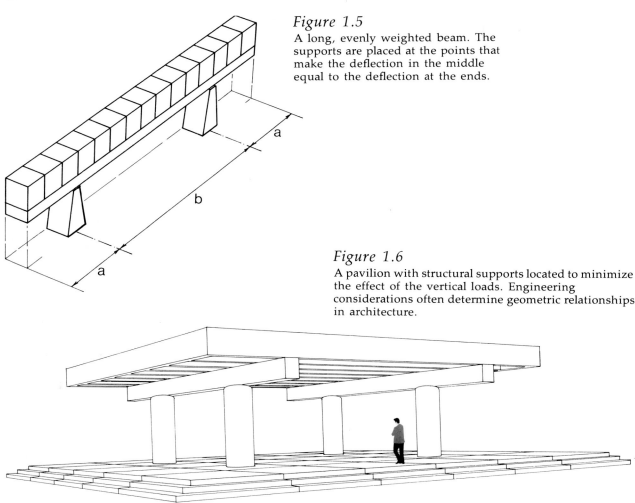

Figure 1.5
A long, evenly weighted beam. The supports are placed at the points that make the deflection in the middle equal to the deflection at the ends.

Figure 1.6
A pavilion with structural supports located to minimize the effect of the vertical loads. Engineering considerations often determine geometric relationships in architecture.

closely spaced roof beams are supported by transverse beams at the points that reduce deflection to a minimum. In turn, the heavier transverse beams are supported by columns at the points that reduce shear to a minimum. The loads are balanced over the supports and the whole arrangement represents an economical use of materials within a rectilinear framework. From the point of view of pure structure, even the most complicated buildings have optimum geometric arrangements of supports which result in the most efficient use of the materials. These ideal structural arrangements often conflict with architectural requirements, however. For instance, if the pavilion in Figure 1.6

were enclosed with glass walls, the columns inside the room probably would interfere with the use of the interior space.

Functional Geometry

Figure 1.7 is a diagram of a Max Bill three-color painting in which no two panels of the same color touch. There are always functional requirements and conditions on the arrangement of architectural spaces similar to the limitations Max Bill put on this painting. Satisfying these conditions within an aesthetic framework is a part of the challenge of architecture.

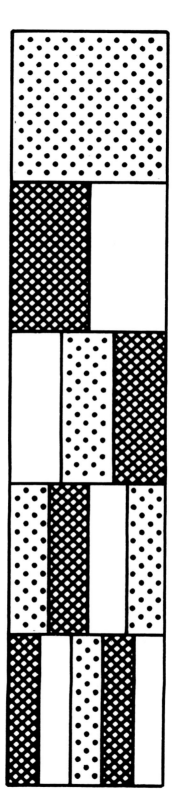

Figure 1.7
Diagram of a Max Bill three-color painting in which no two panels of the same color touch.

Geometry in Architecture

Geometry is everywhere in architecture. Examples from the recent past occur in the work of Le Corbusier (Figure 1.8) and Paul Rudolph (Figure 1.9 and 1.10), where geometric sunshading devices reflecting the movement of the earth with respect

Figure 1.8
An office building in Algiers by Le Corbusier. "This edifice, planned 500 feet in height, has harmony in all its parts. The plan as well as the elevations are implacably controlled by the succession of rational requirements, and expressed with plastic sensibility. The plan is rigorously symmetric, the form a pure imperative prism over which a skin of glass is stretched. Before it a 'cladding' of sun-breakers is applied in a terrace-like form; the direct rays are arrested, rays reflected from the sea are caught, and the beating of rain on the skin of glass is prevented. Over the purity of the silhouette the graduated and diverse repartition of all secondary elements are formed according to the golden mean." (From *Solar Control and Shading Devices* by Aladar Olgyay and Victor Olgyay. Copyright © 1957 by Princeton University Press; reprinted by permission of Princeton University Press.)

Figure 1.9
A guest house on Sanibel Island, Florida, by Paul Rudolph, 1968. In this view the side flaps and curtains are in the closed position to block the rays of the sun. (Photo by Ezra Stoller, © ESTO.)

Figure 1.10
Guest House, Sanibel Island, Florida, by Paul Rudolph. In this view, the side flaps are raised and act as horizontal shading devices, creating an open, airy, and shaded living space. A theme on squares has been carried out throughout the design. (Photo by Ezra Stoller, © ESTO.)

to the sun comprise a dominant element in the exterior design of the buildings. With increasing emphasis on energy conservation, there is certain to be renewed interest in the development of even more sophisticated solutions to problems of sun control, both in the "skin" of the building and in the shape of the building.

Another example of geometry in architecture comes from the sixteenth century: the glorious fan-vaulted roof of King's College Chapel, Cam-bridge, the climax of English Gothic (Figure 1.11). It epitomizes the creation of architectural space through the evolution of intricate geometries.

Geometry in architecture is especially notice-able in roof shapes. Domes, pyramids, half-cylin-ders, cones, prisms, and numerous other shapes are used in the design and construction of roofs. From umbrellas (usually eight-sided polygons) to hyperbolic paraboloids, geometry is an intrinsic part of the design of roofs and shelter.

Figure 1.11
King's College Chapel, Cambridge, England. This chapel was completed in 1515
and represents the climax of English Gothic. (Photo courtesy A. F. Kersting,
London.)

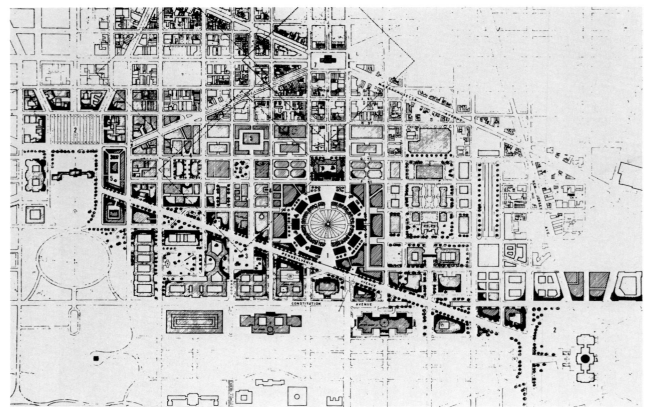

Figure 1.12
A plan for Pennsylvania Avenue in Washington, D.C.

Geometry in City Planning

Figure 1.12 is a 1963 plan for Pennsylvania Avenue in Washington D.C., which consolidated new federal offices in a circular Stonehenge-like arrangement of towers over a rapid transit station midway between the White House and the Capitol. The square space around the Capitol and the rectangular space around the White House have been strengthened by an infilling of buildings, and Pennsylvania Avenue has been restored as an un-interrupted straight-line visual link between the Capitol and the White House, as intended by L'Enfant's original plan.

Clear definition of urban open spaces, the establishment of visual links between important places, the proportion of city blocks, the travel distance across a street grid, and other aspects of city planning involve elements of plane and solid geometry.

CHAPTER 2

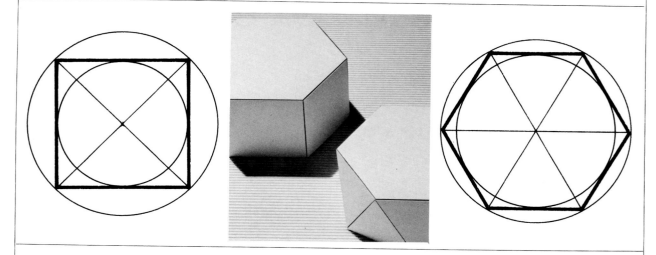

THE REGULAR POLYGONS

Figure 2.1
The Long Gallery of Blickling Hall, Norfolk, England, by Robert Lyminge, 1616.
(Photo courtesy of The National Trust, London.)

THE REGULAR POLYGONS (Figure 2.2) are the elementary shapes of formal geometry. Singly or in combination, they are the basis for prisms, antiprisms, pyramids, and the classical solids as well as numerous plane figures, patterns, and designs important in architecture.

The characteristics of regular polygons are well known. They are symmetrical, equilateral, equiangular; can be inscribed in a circle, and can have any number of sides from two to infinity. When the number of sides becomes very great, a regular polygon becomes a circle, the shape that encloses a given area with the least perimeter.

The straight line is included as a two-sided regular polygon at the beginning because it completes the aesthetic and mathematical requirements of the series. It will later be seen that the circle and the straight line represent the extremes in area enclosure and are perfect complements.

The regular polygons of three, four, and five sides (Figure 2.3) have the same importance in architectural geometry as the primary pigments—red, yellow, and blue—have in color and painting. These shapes—the equilateral triangle, square, and pentagon—and their first truncations—the hexagon, octagon, and decagon—are virtually all that is needed in the beginning pallet of geometric forms and patterns.

TRUNCATION EXPLAINED

Truncation is a word that needs some explanation. In geometry, the term means the changing of one shape into another by altering the corners. The four corners of a square can be progressively cut back to make an octagon; the three corners of an equilateral triangle can be cut back to make a hexagon; in turn, the six corners of a hexagon can be cut back to make a dodecagon; and so on.

Truncation is a process by which a smooth transition with any number of intermediate steps can be made between one shape and another. The process can be imagined as an animated film with any number of frames. The equilateral triangle and hexagon (3/6 series), the square and octagon (4/8 series), and the pentagon and decagon (5/10 series) are primary and secondary polygons linked by the process of truncation. These pairs alone provide the basis for Chapters 3, 4, and 5.

Figure 2.2

Six of the regular polygons. The series begins with a straight line of two sides and ends with the circle.

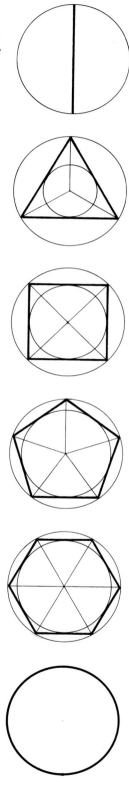

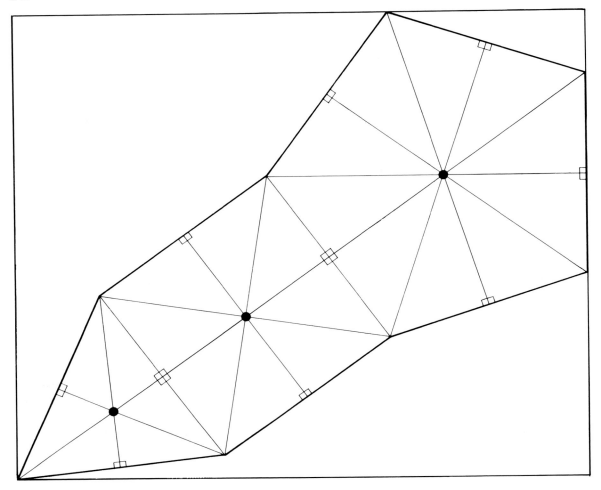

Figure 2.3
The three primary polygons set in a rectangle. There are twenty-four right
triangles in these three polygons.

PARTS OF A POLYGON

Every regular polygon has associated with it a right triangle, an isosceles triangle, and a rectangle. The principal element is the right triangle, and there are twice as many of them as there are sides to the polygon. The height of the right triangle is the inside radius of the polygon, which provides the basis for calculating the area. Back-to-back right triangles form the isosceles triangles of the polygon.

The right triangles that comprise the isosceles triangle can be rearranged to make the rectangle of

the polygon, the diagonal of which is the outside radius of the polygon. Figure 2.4 shows the rectangles of the regular polygons. Thus, for identification purposes, it is possible to speak of the right triangle of the pentagon, the isosceles triangle of the hexagon, the rectangle of the decagon, and completely describe a shape.

The 45° right triangle is both the right triangle and the isosceles triangle of the square, the 30°–60° is the right triangle of both the equilateral triangle and the hexagon, the equilaterial triangle

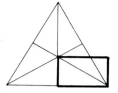

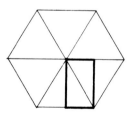

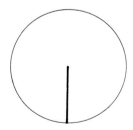

Figure 2.4
The rectangles of the regular polygons. Of these, the rectangle of the pentagon conveys a degree of fullness greater than the others.

is the isosceles triangle of the hexagon. These are very familiar and essential shapes in the geometry of architecture.

POLYGONS IN THE SOLIDS

The part that regular polygons play in plane figures and patterns is well known, but the relationship to the solids is worth reviewing.

When uniform height or thickness is added to a polygon, regular or irregular, it becomes a prism. If the walls are perpendicular to the base, the volume is simply the area of the base times the height. Floor tiles are prisms, most rooms and buildings are rectangular prisms, and even a cylinder (in which the base is the circle of the regular polygon series) is a prism.

If the center of a regular polygon is raised and the isosceles triangles are extended to meet it, the regular polygon becomes a pyramid. Pyramids of the regular polygons, in turn, make up the eighteen classical solids—tetrahedron, cube, and so on. A regular polygon, then, has the potential of becoming a pyramid.

There is yet another series of shapes constructed from the regular polygons, the antiprisms. Antiprisms are curious shapes and probably misnamed since they are neither prisms nor "anti" prisms. They are rarely used. The base of an antiprism is a regular polygon, but the walls are isosceles triangles and are slanted, not perpendicular to the base. Like the classical solids, the internal construction is pyramidal. For comparison, Figure 2.5 shows a regular prism and a regular antiprism.

Prisms, antiprisms, pyramids, and the classical solids are described in later chapters.

PERIMETER LENGTH AND AREA ENCLOSURE

Before unfolding the constructions of the regular polygons, there is a characteristic of plane shapes which is of special importance in architecture. This is the relationship between the area enclosed and the length of perimeter or circumference required to enclose it.

Architecturally, length of perimeter is important because, on the "plus" side, it is the wall area available for doors, windows, picture hanging, acoustical materials, and other features which contribute greatly to the pleasure and enjoyment of a room or space—the more perimeter, the better.

Figure 2.5

A regular prism and a regular antiprism. Regular prisms have square sides and
regular antiprisms have equilateral triangular sides. The tops and bottoms of
both are regular polygons. (Photo by Gerald Ratto.)

On the "minus" side, perimeter is a cost factor related to quantity of materials, maintenance, and heating and cooling loads—the less perimeter, the better within similar types of constructions. As with other aspects of the art, the architect balances conflicting goals in the final solution.

The shortest distance between two points is a straight line, but the shortest distance around an area is a circle. When the distance around is broken into straight-line segments as in the regular polygons, or is itself a wave or a zigzag or some other shape, or is flattened into an ellipse, the distance around an area of constant size will be increased. In this respect, a circle is the most efficient of the plane shapes (a sphere is the most efficient of the solid shapes), and the perimeter efficiencies of other shapes can be measured against it. This has been done, and the results are presented in Figure 2.6, a chart that shows the regular polygons and three other series of shapes

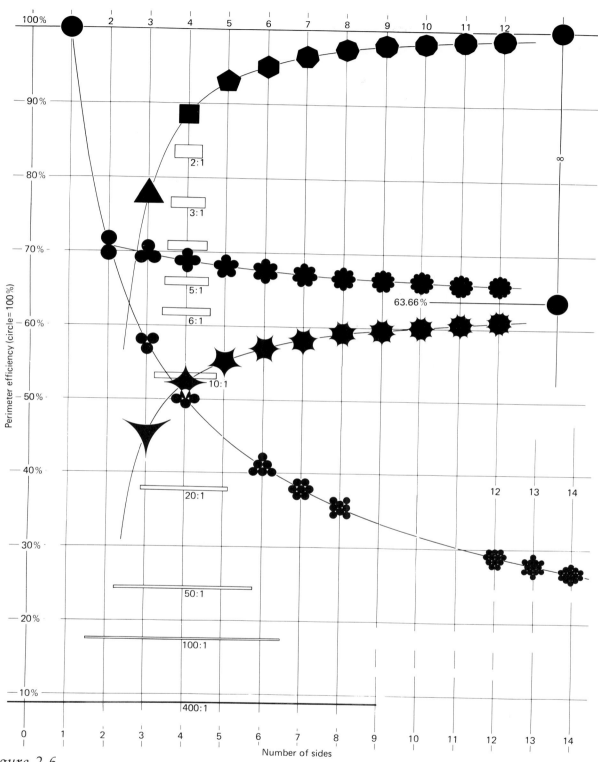

Figure 2.6

A chart of the perimeter efficiency of plane shapes. The shapes nearest the top
have the least perimeter required to enclose a given area. All of the shapes have
the same area although some appear larger than others.

plotted according to their perimeter efficiency and the number of sides in the shape. The shapes closest to the top have the least perimeter. All of the shapes have the same area, only the perimeter changes as a characteristic of shape. For rooms and buildings of the same height, the chart shows the relative amount of wall area provided by the different shapes.

When height is not the same, comparisons of wall area cannot be made from this chart; only the length of perimeter for a given area can be compared.

Figure 2.7 shows the geometric relationship between the four series of shapes seen as solid on the chart.

Two variations of the regular polygons, the concave and convex aspects, sometimes called "cookies" and cusps, are included. The most interesting of these is the three-sided figure in the concave series which, in contrast to a circle, has more than double the perimeter for a given area. The first two or three shapes in the concave series are ideal as columns because they have a large radius of gyration.

When the number of sides becomes very great in both the concave and convex series, the circumference will be made up of tiny semicircles. The shape will look like a circle but the perimeter efficiency will be only 63.7% that of a true circle.

The curve sweeping down from the upper left hand corner of the chart shows the dramatic increase in perimeter that occurs when one shape (in this case, a circle) is divided into two, three, or more identical shapes, the total area remaining the same. Just as splitting a log does not change the quantity of wood but greatly increases the surface area, dividing one shape into smaller shapes greatly increases the perimeter and hence opportunity for daylight, views, and so on. A two-story building will almost always have more outside wall area (and, of course, less roof and foundation area) than a one-story building with the same total floor area. One of the great virtues of the traditional two-story rectangular house is that as many as seven or eight rooms will have at least two different outside exposures. Multistory buildings, which are identical floor plans stacked one above the other, always have a great deal of perimeter with respect to the area enclosed.

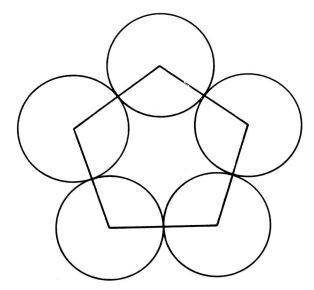

Figure 2.7
Five circles, a pentagon, and the concave and convex aspects of the pentagon combined in one drawing. These are the four series of shapes shown on the chart of perimeter efficiency.

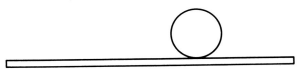

Figure 2.8
A circle and a long rectangle with the same area. The perimeter of the rectangle is 4.5 times that of the circle.

THE EXTREMES IN AREA ENCLOSURE

Rectangles on the chart are shown only in outline form because they are developed in detail in Chapter 7. For now, one feature should be observed. In contrast to the circle, very long rectangles have an immense amount of perimeter with respect to the area enclosed. They have the lowest efficiency of any of the shapes shown on the chart. Thus a circle and a long rectangle are the extremes in area enclosure.

Figure 2.8 illustrates a circle and a long rectangle with exactly the same area. The perimeter of the rectangle is 4.5 times that of the circle!

An architectural example of this contrast in

Figure 2.9
Chapel at M.I.T. by Eero Saarinen, 1955. The long rectangular narthex is in marked contrast to the circular brick chapel but the whole remains harmonious. (Photo by Joseph W. Molitor.)

geometric shape is found in the nondenominational chapel at M.I.T., designed by Eero Saarinen (Figure 2.9). The dimly lighted cylindrical brick chapel is entered through a long rectangular glass-walled narthex. Between the narthex and the chapel there is a difference in perimeter efficiency of nearly 25 per cent. Considering the range of shapes normally used by architects, this extremely strong contrast is enhanced by equally strong contrasts in materials, brightness, and height.

The shape with the least perimeter is by no means always the best. A painter, for instance, who might want a maximum of wall surface on which to hang paintings, would do well with a long rectangular room (or two square rooms), but a circle or a square would be a poor choice.

There is, incidentally, only about a 5% difference in perimeter length between a square room and one with a length-to-width ratio of 2:1, the maximum for most rooms. To increase perimeter decisively, the shape must be exaggerated. This has been done in the magnificent Long Gallery of Bickling Hall shown at the beginning of the chapter (Figure 2.1).

CHAPTER 3

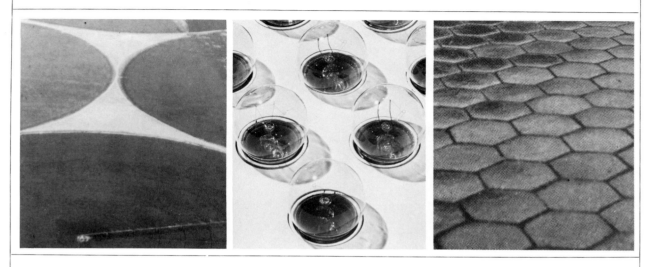

EQUILATERAL TRIANGLE AND HEXAGON

Figure 3.1
A Hungarian thatched hut based on the equilateral
triangle. (Photo by Werner Bischof, courtesy of
Magnum Photos, Inc.)

THE EQUILATERAL TRIANGLE is the first of the geometric primaries. It holds the same position and importance in geometry as the color red occupies in the color primaries of red, yellow, and blue. The hexagon is its truncated version, a geometric secondary. Equilateral triangles and hexagons make up the 3/6 series of shapes.

The equilateral triangle is an isosceles triangle. Isosceles triangles (Figure 3.2) are not to be underrated. They are parts of the regular polygons, the sides of pyramids and antiprisms, and the interior partitions of the classical solids. In architecture, they are seen as roof slopes, low for rain, steep for snow. A-frame dwellings, ancient and modern (Figures 3.1 and 3.3), are prime examples. The mid point of the series of isosceles triangles is the isosceles triangle of the square with an apex of 90°, the 45° right triangle.

Figure 3.3
The Flender House, an early A-frame in Stowe, Vermont, by Henrik Bull, 1953. (Photo by the architect.)

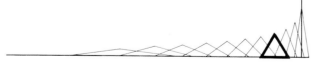

Figure 3.2
The series of isosceles triangles with apex angles from 0° to 180°. The equilateral triangle is just one in the series, but it has unique attributes.

THE EQUILATERAL TRIANGLE

Of the isosceles triangles, the equilateral triangle (Figure 3.4) is the most compact, the most versatile, and the most frequently used. Within the 3/6 series of shapes, the 60° grid comprised of equilaterals is as important as the universal square grid is in the rectilinear framework.

The right triangle of the equilateral is the 30°–60°, used for a multitude of patterns, grids, and designs, often with the illusion of depth (Figures 3.5 and 3.6). The rectangle of the equilateral triangle is slightly more elongated than the golden mean. It has a length-to-width ratio equal to the square root of three. In fact, all of the important dimensions of the equilateral triangle—the height, outside radius, area, and inside radius—

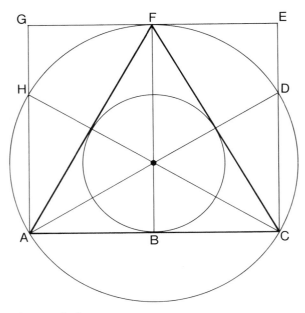

Figure 3.4
The equilateral triangle.

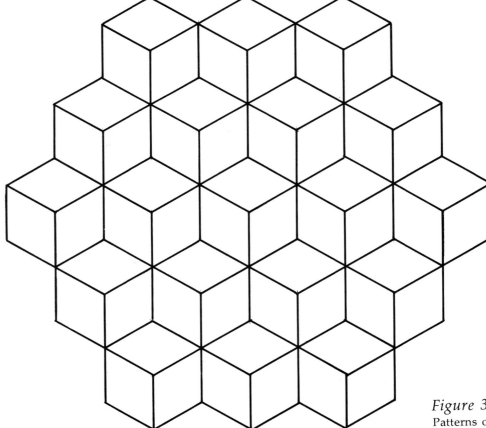

Figure 3.5
Patterns of the 30°–60° right triangle.

Figure 3.6
Patterns of the 30°–60° right triangle.

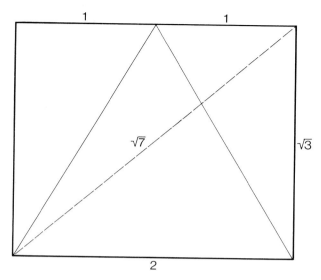

Figure 3.7
The rectangle enclosing the equilateral triangle. This rectangle is used for patterns based on the 60° grid.

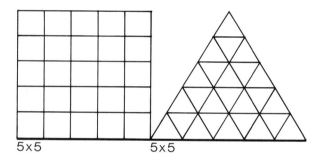

Figure 3.8
The similar subdivision of an equilateral triangle and square.

are functions of the square root of three. For centuries, people have found the equilateral triangle to be spiritually significant and a powerful symbol. If there is a mystique associated with the equilateral triangle, it must also be the mystique of the square root of three.

The rectangle that surrounds the equilateral triangle (Figure 3.7) is also noteworthy. It is just slightly off-square in proportion and fits all of the patterns based on the 60° grid. The dimensions of this rectangle embody the square root of three, and the diagonal includes as well the square root of seven, one of only two or three instances where this number occurs in geometry.

There are seemingly endless variations on the rectangle of the equilateral triangle, the equilateral triangle itself, and the hexagon. The common denominator is the square root of three, which, like the key of C minor in a symphony, can be the principal tonality in an architectural composition. The mathematical relationships will be in harmony, that is a certainty, and the visual images may be as well.

Equilateral triangles comprise the surface of three of the five Platonic solids—tetra, octa, and icosa—and they occur as secondary faces in no less than nine of the thirteen semiregular solids. The walls of regular antiprisms and the walls of the five "perfect" pyramids are equilaterals.

In common with a square, the number of panels in the subdivision of an equilateral triangle is the square of the number of divisions in a side (Figure 3.8). The surface of any equilateral can be constructed entirely of modular equilaterals.

The grid of the equilateral triangle, the 60° grid (Figure 3.9), is the geometric basis for many patterns including the well-known hexagonal pattern of tiles, the grouping of circles, and the beautiful flat patterns associated with the great and small rhombicosi (Chapter 13).

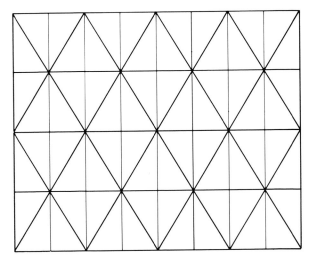

Figure 3.9
The 60° grid and the rectangles of the equilateral.

Equilaterals, and all triangles, are rigid shapes. When shapes are cut from cardboard this aspect is of no significance, but when the shapes are framed in steel, aluminum, or wood, rigidity is important. The simplest connection of the framing members is a pin, which permits rotation. When pin connections are used, an equilateral is rigid, but a square, pentagon, or other polygon will collapse under pressure from the side. For this reason, equilateral and other triangles frequently are seen in bridges and trusses and as an alternative method of bracing in building. An unusually dominant expression of this bracing is found in the Alcoa office building in San Francisco (Figure 3.10).

Figure 3.10
The Alcoa Office Building in San Francisco by Skidmore, Owings and Merrill, 1963. (Photo © by Morley Baer.)

HEXAGONS AND THE HONEYCOMB

The equilateral triangle is the isosceles triangle of the hexagon. The close relationship between the two shapes can be seen in the hexagonal plan of Granmichele, a city of about 15,000 in northern Sicily (Figure 3.11). Hexagons are also the shape of snowflakes and they are the shape of textured tiles on the sidewalks of Fifth Avenue, New York (Figures 3.12 and 3.13). Frank Lloyd Wright and other architects have used hexagonal themes for the design of entire buildings and all the furniture therein. A grid of small hexagons, which is itself constructed over the grid of the equilateral triangle, provides the geometric basis for the pentagonal pieces in the flower pattern illustrated in Figure 3.14.

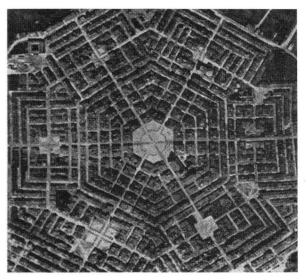

Figure 3.11

An aerial view of Granmichele, a hexagonal city in Sicily. (From *The Nature of Cities* by L. Hilberseimer, © 1955 by Paul Theobald—all rights reserved.)

Figure 3.12

Precast hexagonal sidewalk on Fifth Avenue in New York City. (From *Cities* by Lawrence Halprin, © 1963 Reinhold Publishing, reprinted by permission of Van Nostrand Reinhold Company. Photo by Lawrence Halprin.)

Figure 3.13
Sidewalk, Fifth Avenue, New York. (From *Cities* by Lawrence Halprin, © 1963 Reinhold Publishing, reprinted by permission of Van Nostrand Reinhold Company. Photo by Lawrence Halprin.)

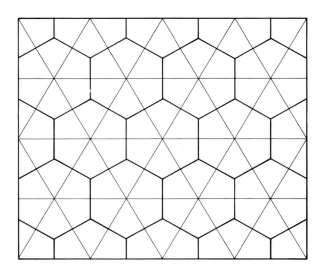

Figure 3.15
The standard hexagonal grid.

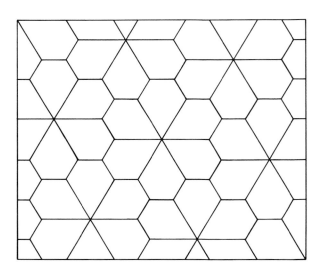

Figure 3.14
Pentagonal pieces in a hexagonal pattern.

Figure 3.15 is the standard hexagonal grid, and Figure 3.16 is a close-up of the cells of a comb from a hornet's nest, nature's version of the hexagonal grid. The comb is constructed from wood pulp and designed primarily for protection—to shield the eggs and larvae of the hornet from rain, cold, and heat. The honeycomb is one of nature's most ingenious structures, compact and economical, the perfect environment for embryonic hornets. It is a repetition of identical cells in a hexagonal pattern.

Figure 3.16
The cells of a comb from a hornet's nest. Photo by Andreas Feininger. (Life Magazine © Time Inc.)

While ideally suited to incubating hornets, the cellular concept has serious drawbacks for humans. It is inflexible, cannot accommodate a variety of space requirements, and offers only limited opportunity for daylight and ventilation.

A school, for example, may have thirty classrooms with nearly identical requirements. The classrooms could all be hexagonal in shape, although that might not be the best choice, and they could be arranged in a honeycomb pattern. But to these thirty classrooms must be added a gymnasium, cafeteria, auditorium, principal's office, corridors, and other diverse spaces that do not fit into a pattern of repetitive shapes. This is one reason why architects prefer the rectangular framework and develop a fondness for the aesthetics of it. The rectangular grid can accommodate a variety of spaces, large and small, in a cohesive whole without waste of space.

As another drawback, the cells of a honeycomb have no exposed perimeter and little possibility for the penetration of sunlight. A skillful architect uses daylight, just as he or she uses shape, structure, color, texture, and materials, to infuse space with special vibrancy. Indeed, if asked to select only one from the list of important architectural ingredients, many architects would choose daylight. The whole study of perimeter is based largely on the importance of daylight—as well as natural ventilation and views—for human habitation. While the concept of repetition is indispensable for economy, the cellular arrangement of repetitive units rarely is acceptable for spaces inhabited by humans.

THE GROUPING OF CIRCLES

The grouping of circles (Figure 3.17) follows the hexagonal pattern and is the ideal pattern for an orchard. Although not as practical as straight rows, it is the one arrangement that gives the maximum opportunity for each tree to grow and flourish. The pattern of the grouping of circles caused by huge circular irrigation systems (Figure 3.18) which cover as much as 16 acres with one sweep of the arm can also be seen on the earth. The thirteen-bulb lamp (Figure 3.19) is based on the same pattern and permits at least a dozen different geometric arrangements depending on which bulbs are illuminated.

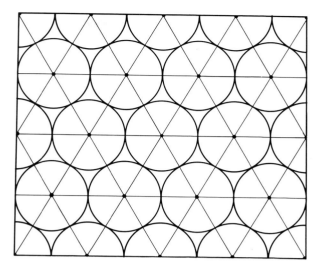

Figure 3.17
The grouping of circles on the 60° grid.

THE 30°–60° LINKED PATTERNS

Just as one solid shape leads to another through the process of truncation, one plane pattern can lead to another. Five of the 30°–60° tile patterns are linked in this manner (Figure 3.20).

Truncation of the hexagon of the honeycomb pattern (upper left) results in the dodecagons of the fragile 12–3 pattern in the upper right corner. If truncation continues in the same direction, the dodecagons become the hexagons of the 6–3 pattern seen at the center of the figure.

Similarly, the triangles of the equilateral grid (lower right) can be truncated to make the white hexagons of the familiar 6–6 floor pattern in the lower left corner. The black hexagons in this pattern occur at the interstices of the original pattern. If truncation continues from the same corners where it began, the white hexagons of the 6–6 pattern become the triangular pieces in the 6–3 pattern at the center.

The diagram represents a continuous process of transformation beginning at either end and culminating in the 6–3 pattern at the center.

Each of these flat patterns has a three-dimensional counterpart in the icosa-dodeca series of the classical solids (Chapter 13).

Figure 3.18
Aerial view of center-pivot self-propelled irrigation sprinklers.

Figure 3.19
A thirteen-bulb lamp based on the grouping of circles. (Photo by Gerald Ratto.)

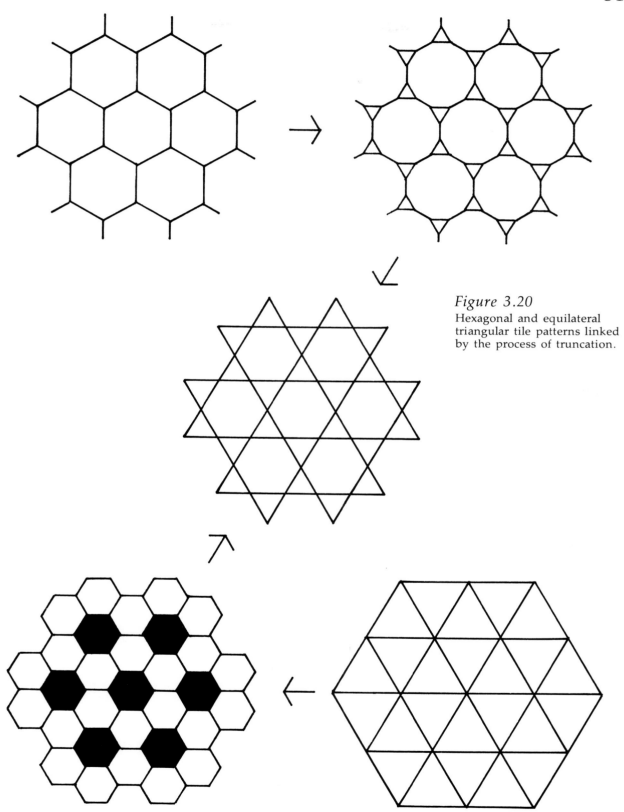

Figure 3.20
Hexagonal and equilateral triangular tile patterns linked by the process of truncation.

C H A P T E R 4

SQUARE, OCTAGON, AND THE PROGRESSION OF FOURTHS

THE SQUARE is probably used more frequently than any other shape in architecture. Most major buildings are square in plan, have square bays, or use a square grid system. Major open spaces, especially those indispensable ones at the heart of urban activity, often are squares. Most of the United States is crisscrossed by a square grid at one-mile intervals.

Octagons, which are linked to squares by the process of truncation, are rare today. Octagonal houses were once popular because of the development and promotion of a patented wood framing system. Many cities have one. Orthodox churches frequently have octagonal interiors which make a lasting visual impression. The square and the octagon have been used as symbols throughout history. The cabalistic sign (Figure 4.1) is one example. Of course, there are many beautiful decorative tile patterns based on the square and octagon. One of these is shown in Figure 4.2.

In 1457 Antonio Filarete, a Florentine architect and sculptor who built the bronze doors of St. Peter's in Rome, designed a star-shaped city called Sforzinda based on the juxtaposition of two squares (Figure 4.3). The outer walls formed a sixteen-sided figure. His central square, incidentally, is twice as long as it is wide, a 2:1 rectangle. The sixteen-sided figure does not occur again but the square and the eight-pointed octagon occur today in city planning as diamond and diagonal shapes (Chapter 8).

GEOMETRIES OF THE SQUARE AND THE OCTAGON

The square is a geometric primary, corresponding to yellow in the color primaries. Its truncated version is the octagon, a geometric secondary.

Squares are the walls of regular prisms, the surfaces of cubes, and the bases of the important

Figure 4.1

A cabalistic sign on the main nave wall of a mosque at Kairouan, Tunisia. (From *The Matrix of Man* by Sibyl Moholy-Nagy © 1968 Frederick Praeger.)

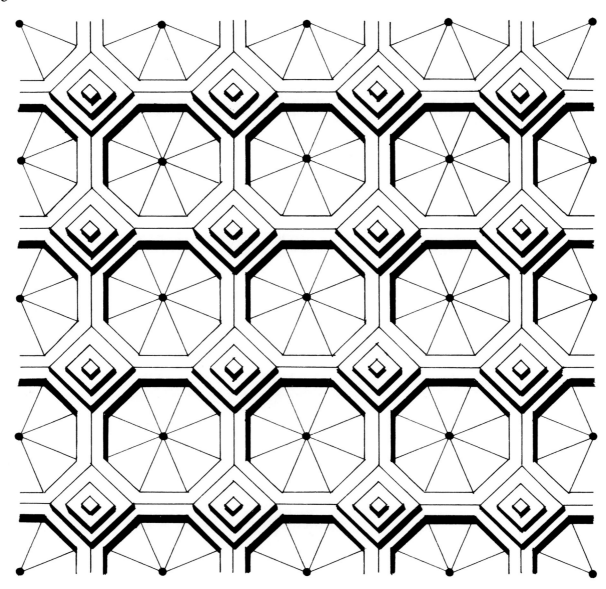

Figure 4.2
Square and octagon tile design.

four-sided pyramids. They appear on the surface of seven of the thirteen semiregular solids. Octagons appear on only two.

The square is a rectangle. It stands at the center of the universe of rectangles (Figure 4.4). Although it is the most efficient in terms of perimeter it is proportionless and thus not always the best rectangle for many uses. Rectangles as room shapes and as building shapes using a square grid are the subject of Chapter 7. Rectangles, it should be noted, are simply flattened squares.

Just as the square root of three is associated with the equilateral triangle and hexagon, the square root of two, the length of the diagonal of a square with sides equal to one, is the number constantly recurring in square and octagonal geometries.

THE SQUARE GRID

In 1785 the Congress of the United States passed

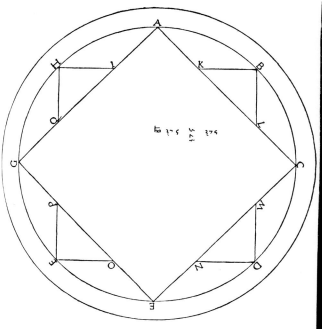

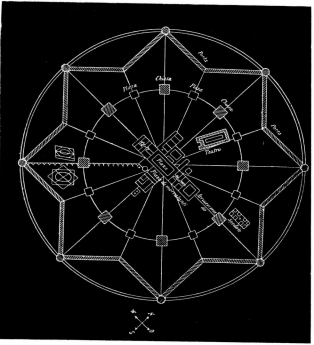

Figure 4.3
Sforzinda, the eight-pointed, star-shaped city by the architect Filarete, 1457. The shape is of two squares superimposed diagonally within a circle. (From *Space, Time and Architecture* by Siegfreid Giedion, Harvard University Press, 1967; reprinted by permission of the publishers.)

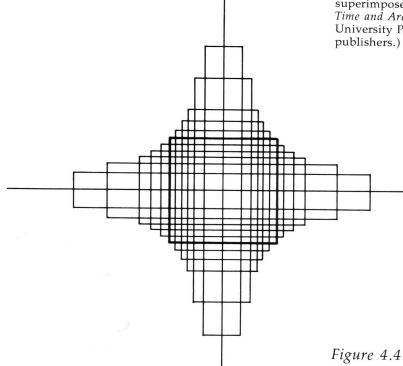

Figure 4.4
The array of rectangles. The square is at the center and is the most efficient in terms of perimeter but it is not the best proportion for all uses.

an ordinance establishing a uniform north–south, east–west reference grid over virtually the entire country. This grid divided the land on paper into square-mile parcels (640 acres). It has been used ever since to establish and maintain a degree of order in land use and subdivision. The influence of this grid on the landscape can be seen from the air in many parts of the country by the rectangular arrangement of fencelines and roads, by the pattern of crop plantings and land parcelization, and by the boundaries of housing subdivisions. Although the grid was drawn without regard to topography, it is nonetheless a useful system which permits expansions and contractions in land use over time on a well-defined framework. The system plainly does not inhibit any land use activity but it may, at times, impose unnatural boundary lines on that activity.

Cities and townships use this U.S. grid or establish one of their own, but invariably a planning grid of some kind keeps the individual parcels of land organized within the rectilinear framework as changes occur over time. Buildings, in turn, fit on the parcels established by the system of property subdivision.

The story of the grid does not end at the property line, however. Major buildings usually have a square grid of their own. This grid permits structural lines, partitioning, windows, ceiling panels, lighting, and other systems to be dimensionally coordinated and arranged so that one does not interfere with the other. The square grid system, which is almost universal, permits change in one part or system without a "ripple" effect throughout every other part or system. Within limits, it

enables a person or team to concentrate on one aspect of the building design while others proceed independently on other aspects. In the end, if all goes well, the parts and pieces and systems from many different sources will fit together into a cohesive whole. The square grid then is the building block for much of architecture. Like the grid established by Congress on the landscape, it may or may not be visible in the final product.

TRUNCATION PATTERNS OF THE SQUARE GRID

Truncation of the squares in a square grid (Figure 4.5) results in a pattern of octagons with an equal number of small squares in the leftover spaces (shown in black), the pattern seen in the middle. If truncation continues in the same direction, the pattern in the middle becomes one of alternate black and white squares turned diagonally across the squares of the original grid, the pattern on the right. This last illustration is the classic tile pattern used on the floors of St. Paul's Cathedral, Westminister Abbey, the rotunda of the California State Capitol (Figure 4.6), and countless other rooms and places. It is a checkerboard pattern but a checkerboard pattern turned at 45° with respect to the walls of the room.

These flat patterns of the square grid have three-dimensional counterparts beginning with the cube in the series of classical solids (Chapter 13).

The sequence of successive truncation of the square itself is one of four-, eight-, sixteen-, and

Figure 4.5
Patterns resulting from the successive truncation of the squares in a square grid. The checkerboard pattern is turned at 45° with respect to the enclosing square.

Figure 4.6
The rotunda of the California State Capitol as recently remodeled by Welton Becket and Associates. (Photo by Marvin Rand.)

thirty-two-sided regular polygons. The number of sides doubles with each successive truncation. These polygons, and, in fact, all of the regular polygons with a number of sides evenly divisible by four, have sides parallel to the right-angle Cartesian coordinate system, the standard grid system. They are therefore especially useful in architecture, and they can be grouped together in a series called the progression of fourths.

THE PROGRESSION OF FOURTHS

The natural evolution of geometric patterns and shapes appeals to every architect. As mathematical artists, they are intrigued with the infinite structural possibilities offered by simple lines and basic geometric shapes, while as functional designers, their daily concern is the space enclosed by lines and shapes—the aesthetics, human dimensions of scale, and the most efficient and effective use of materials.

The primary example of geometric evolution is the progression of regular polygons from a line to an equilateral triangle, square, pentagon, hexagon, and on to infinity. As the number of sides increases, the polygon becomes more and more like a circle, until at some point the sides become so short that the shape, in fact, becomes indistinguishable from a circle. As long as it has sides and distinct lines, however, the shape remains in the polygon family (Figure 4.7).

An architectural example of a circlelike polygon is the Wells Fargo Bank in downtown San Fran-

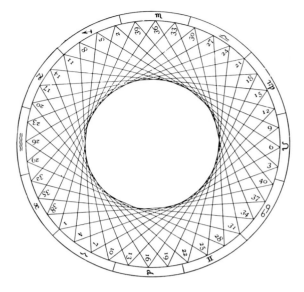

Figure 4.7
A regular polygon of forty sides constructed by a continuous string. (From *Pioneers of Science* by Sir Oliver Lodge, Macmillan & Co., London, 1908.)

cisco (Figure 4.8). This forty-sided building closely approximates a circle and is referred to as the "round bank" although it is constructed entirely of straight-line elements.

The fact that the bank has forty sides is no accident of design. Of all the shapes in the regular polygon series, those that have a number of sides evenly divisible by four are exceptionally suited to architecture. They unfailingly relate to the Cartesian grid system—the rectilinear framework within which most of our constructed objects fit

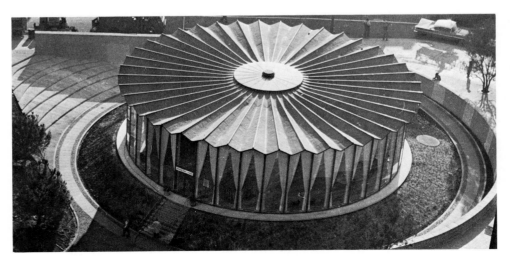

Figure 4.8
The Wells Fargo Bank, Crown Zellerbach Plaza, San Francisco, by Hertzka & Knowles and Skidmore, Owings and Merrill, 1958. (Photo by Jon Brenneis.)

and relate to each other, whether they be rooms, buildings, streets, or cities.

If, in the series of regular polygons, only the fourths are used, the shapes with a number of sides evenly divisible by four, the series begins with a square, not with a line. It progresses to an octagon, a dodecagon, and so on, in multiples of four. These shapes can be combined in groups of four to form a larger shape which will have lines on all sides in common with an enclosing square (Figure 4.9).

As the number of sides increases, the lines in common become shorter and shorter, approaching a single point of tangency. No matter how small, however, they will still hold the shape, even a circle, securely in the Cartesian grid system.

When the shapes in this "progression of fourths" are combined into larger groups of four,

Figure 4.9
Assemblies of the regular polygons with a number of sides divisible by four.

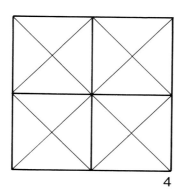

4

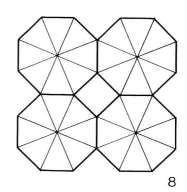

8

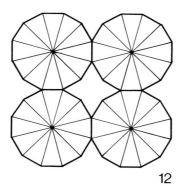

12

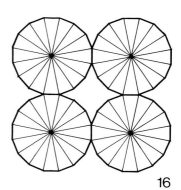

16

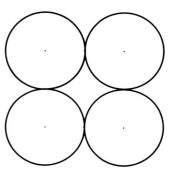

the space at the center takes on new interest. While squares do not create a new shape when assembled, octagons in combination produce a small diamond at the center, and dodecagons form a four-pointed star. According to the *Book of Signs* (a Dover book by Rudolf Koch), the phenomenon of this cruciform shape carries a grave and solemn warning. The "fat" star produced at the center of the dodecagons illustrated here, however, hardly seems ominous.

As the number of sides become greater, the central space formed by four circles, actually near-circles, becomes simply the concave aspect of the square.

If the outlines of the polygons are removed from the groups of four seen in Figure 4.9, a concentration of lines emerges which is visual dynamics of the first order (Figure 4.10). The sequence, developed in multiples of four, always retains a parallel relationship to the square and to the universal

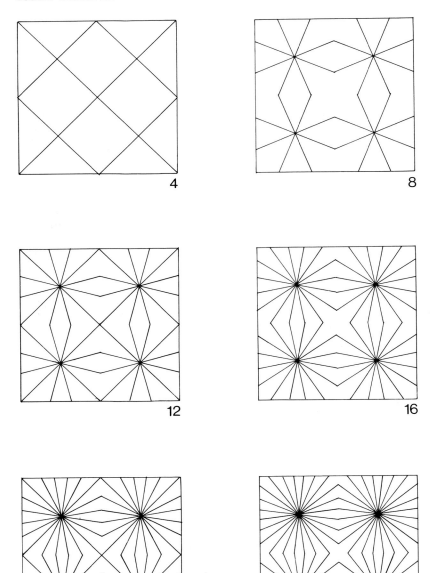

Figure 4.10
The patterns created by the assembly of polygons in the progression of fourths.

grid. Note the recurrence of the four-pointed star at the center, now more presaging than the earlier form.

SUCCESSIVE APPROXIMATIONS TO THE SPHERE

The same evolutionary process of polygons, viewed three-dimensionally, produces a dramatic series of shapes (Figure 4.11) that successively approximate the sphere. Just as polygons can be combined in groups of four, these solids can be combined into groups of eight, always touching the faces of an imaginary enclosing cube at twenty-four points—the quarter points of the six faces.

As in the plane shapes, this progression of fourths series in solid form depends on regular polygons with a number of sides evenly divisible by four—square, octagon, dodecagon, and so on. The equatorial belt of each of these convex polyhedra is this regular polygon, and a cross section through the center of each of the shapes will be an identical polygon.

This series begins with a cube and progresses to

Figure 4.11
Successive approximations to the sphere by the progression of fourths. (Photo by Colin McRae.)

an eight-sided figure in plan and section, a twelve-sided figure, a sixteen-sided figure, and so on to infinity.

As the number of sides increases, the shape becomes closer and closer to a sphere, and at some point becomes visually indistinguishable from a sphere.

The architectural character of this series of fourths is notable:

1. The horizontal plan at the equator and the vertical cross section through the middle are always identical.

2. One dimension, the width, is exactly the same for every piece in the construction, which means that the pieces can be cut or fabricated from a roll of material of constant width. The whole spherical shape can be covered without waste.

3. The plan at every horizontal level is the same shape, a regular polygon.

4. The center of each panel touches an imaginary enclosed sphere.

This series provides a way to approximate a sphere to whatever degree desired with minimal connections, calculations, and cuts. In terms of surface efficiency, the progression is from 80.6% (four sides), to 95.6% (eight sides), to 98.1% (twelve sides) to 98.9% (sixteen sides), and so on to the 100% efficiency of the sphere.

The progression of fourths series satisfies Euler's formula, which states that the number of vertices plus the number of faces equals the number of edges plus two.

The two-dimensional equivalent to the successive approximations (Figure 4.12) is nearly identical to the spider web (Figure 4.13).

Spheres developed from this series pack together on a regular square Cartesian grid, which differs from the nesting pattern usually assumed in the close-packing of spheres. Figure 4.14 shows on the left a group of eight tangent spheres with their centers being the centers of close-packed cubes. The equilibrium—the balance of stresses—although delicate, is stable.

The grouping on the right is the traditional way to assemble spheres, with each nesting in the other, a group of twelve spheres around a center sphere, which can be repeated indefinitely. The

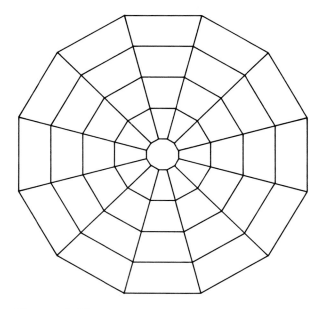

Figure 4.12
The flat pattern of the successive approximations to the sphere. The four-way intersections are typical of the spider web and other woven patterns.

centers of the outside spheres form a cuboctahedron, the geometric solid of six square faces and eight trianglar faces, described in Chapter 13.

This is the natural method of piling round objects such as shot or cannonballs together and often is assumed to be the most stable. From an architectural point of view, however, which references nearly all construction to the rectangular framework, the progression of fourths method is equally important. The point of contact is minute, but if the "glue" is strong enough, the essential stability will endure.

SUCCESSIVE APPROXIMATIONS TO THE ELLIPSE

The selective use of polygons, beginning with the square and progressing in multiples of four, leads to closer and closer approximations of the circle and to the sphere within the rectilinear framework.

There is still another shape which can be approximated with the magical formula of fourths, and that is the ellipse.

As a shape, the ellipse combines the appeal of the circle with the advantage of an axis, that is, a

Figure 4.13
Orb web made by the common garden spider. Photo by Andreas Feininger. (Life Magazine © Time Inc.)

Figure 4.14
Cube and cubocta arrangements of spheres. When construction is based on the progression of fourths, the cube arrangement is stable. (Photo by Gerald Ratto.)

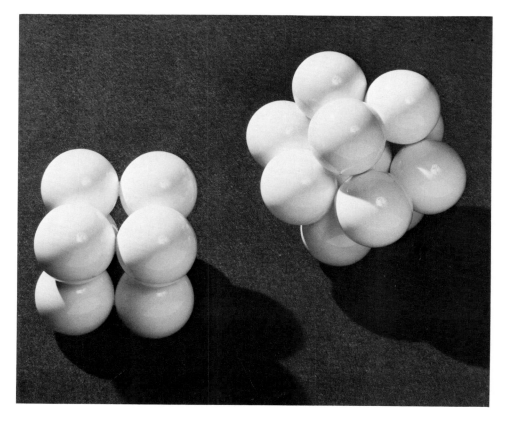

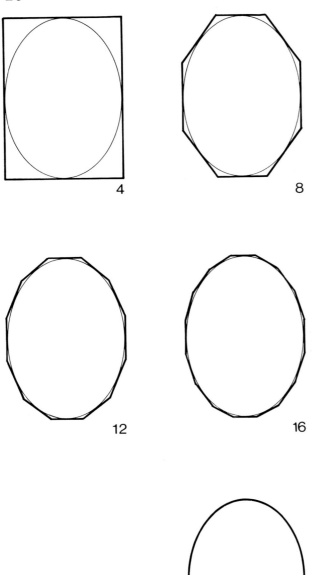

definite north–south line and a definite east–west line, a major and a minor axis, which clearly establishes direction and orientation within the enclosure.

From an architect's viewpoint, the ellipse has two problems. First, the formula for the circumference of an ellipse is complicated, and to an architect's way of thinking, inexact. It is a myth, of course, that the formula is inexact but it lacks the elementary "length plus width times two" logic which 99% of the time yields a dependable and exact answer.

The second and more serious problem is that the curve of an ellipse is continuously changing, which means that it cannot be built simply with straight-line materials conventionally used.

Both problems are solved, however, when the ellipse, like the circle, is broken down into straight-line segments, resulting in a shape that is an approximation but in many instances equally pleasing.

An ellipse is the shape that results from cutting a cylinder at an angle. The 45° ellipse results from cutting a cylinder at 45 degrees. Approximations to the ellipse can be made by cutting regular prisms at an angle, commencing with the square prism, an octagonal prism, and so on. Approximations to the 45° ellipse by this method are shown in Figure 4.15. These approximations will lack the essential Cartesian relationship unless the number of sides in the prisms is evenly divisible by four. And so this series, too, properly belongs with the progression of fourths.

Figure 4.15
The successive approximations to a 45° ellipse.

CHAPTER 5

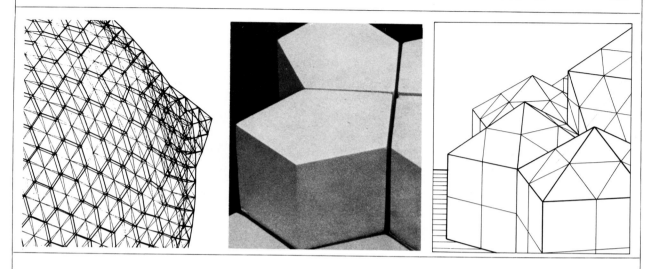

PENTAGON
AND DECAGON

Figure 5.1
I Saw the Figure 5 in Gold, a painting by Charles Demuth, 1928. (From the Alfred
Stieglitz Collection of The Metropolitan Museum of Art, New York.)

Since humans began to count, the number five has had special significance and meaning (Figure 5.1), mainly because of the number of fingers on the hand. Ten is the number considered perfect by the ancient Greek mathematicians and philosophers.

Geometrically, shapes of five and ten sides, the pentagon and decagon, are not less important. The pentagon is another of the geometric primaries, corresponding to blue in the color primaries. The ten-sided decagon is its truncated version, a geometric secondary. Pentagon and decagon are shown together in Figure 5.2.

Both the pentagon and the decagon fall outside of the progression of fourths. Neither has sides parallel to the Cartesian grid system. Three-dimensional constructions of pentagons and decagons, the icosadodeca series of solid shapes, also have a plane system different from the Cartesian rectilinear system of right angled walls and floors.

FIGURES OF THE PENTAGON

The number of importance in the mathematics of the 5/10 series is 1.618, one-half plus one-half of the square root of five ($\frac{1}{2} + \frac{1}{2}\sqrt{5}$), designated by the Greek letter ϕ. Because of the nature of its reciprocal, square, and so on, ϕ is a unique number. It is also the length-to-width ratio of the rectangle of the golden mean. This number, or its reciprocal, occurs in the dimensions of five- and ten-sided figures, both plane and solid. The "sublime" triangle, the isosceles triangle of the decagon with an apex angle of 36° (Figure 5.3), involves ϕ.

Shapes with a common mathematical base such as ϕ or the square root of three have visual similarities. A successful architectural space depends on visual echo, reverberation, and emphasis to develop a theme to its fullest dimensions. In this regard, the visual harmonics between different shapes used for different purposes are an essential part of design.

The rectangle of the pentagon (Figure 5.4) with a length-to-width ratio of 1.376:1 is less extreme in proportion than the golden mean and is invariably

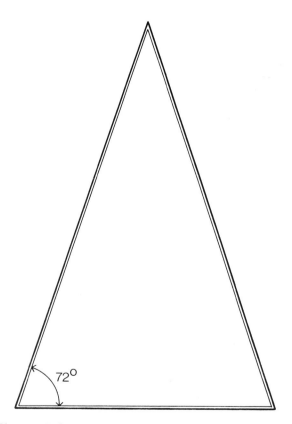

Figure 5.2
Five decagons centered on the points of a pentagon.

Figure 5.3
The "sublime" triangle with an apex angle of 36°. This is the isosceles triangle of the decagon.

Figure 5.4
The rectangle of the pentagon.

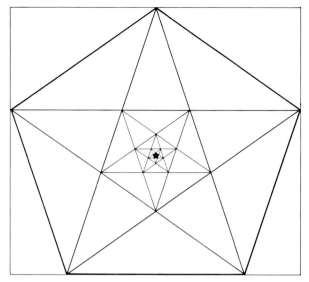

Figure 5.5
Rectangle, pentagon, and five-pointed stars.

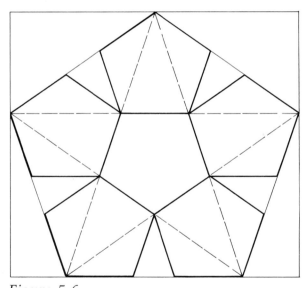

Figure 5.6
The pattern for one-half of a dodecahedron.

successful as a frame for a view, a portrait, or a mirror.

Figure 5.5 shows a progression of star shapes within pentagons. The rectangle that encloses the pentagon, like the rectangle that encloses the equilateral triangle, is slightly off-square. The points of the stars are "sublime" triangles and, when folded up, make a beautiful pyramid, one of the pyramids of the icosadodeca, another of the classical solids.

Figure 5.6 shows five pentagons arranged around a sixth in a flowerlike pattern which can be folded to make half of a dodecahedron. There are, incidentally, always five petals in an apple blossom, an example from nature of this pentagonal display.

Twelve pentagons seamed at the sides make a whole dodecahedron, a Platonic solid corresponding in space to the standard flat pattern of hexagons on a plane. When pentagons or decagons occur in a classical solid, the number of them is always twelve, never more and never less.

Pentagons have an odd number of sides and are not in the progression of fourths, but they are the first of the regular polygons which begin to resemble a circle and have the characteristic compactness of a circle. In fact, of the prismatic solids with square sides, the pentagon is the most efficient volume enclosure when all surfaces are counted, more efficient even than the cube.

Even so, pentagonal rooms are rare, although the shape has appealing features. The corners are

Figure 5.7
The Pentagon. (U.S. Air Force Photo.)

more open than those of rectangular rooms, the walls are not parallel, and there is a strong focus on the axis at the center.

THE STARFISH AND PENTAGONAL TOWERS

Pentagons occur in nature, although infrequently. The pentagonal structure of starfish and sand dollars is a form suited to the sedentary habits of these creatures which must continually be pre-

pared to meet environmental threats to their safety on all sides. The architectural parallel is unmistakable since buildings (unlike space ships, airplanes, trains, ships, and cars) are stationary and must confront the environment on all sides. The prismatic effect of pentagons, especially in a tall building, could be quite beautiful.

The Pentagon, headquarters of the U.S. Department of Defense near Washington, D.C. (Figure 5.7), is the best known constructed pentagonal shape in the world. It is also one of the world's largest office buildings, built low during World

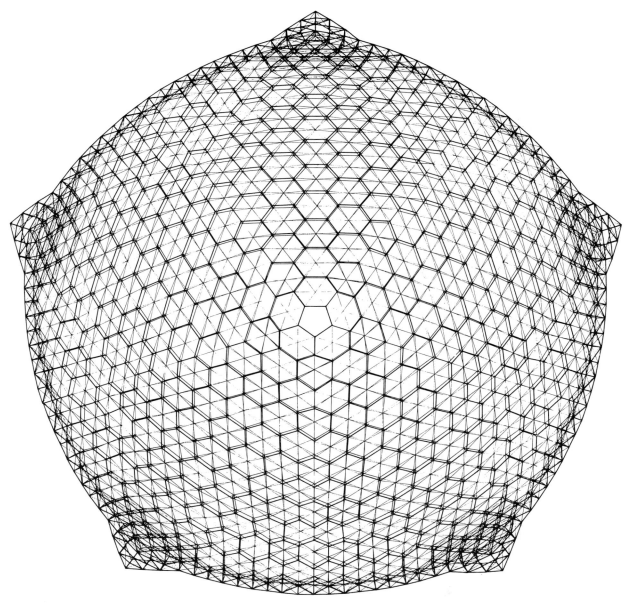

Figure 5.8
The plan of the dome for the American Society for Metals, Cleveland, Ohio, by
Synergetics Incorporated. The pieces are hexagons but the shape is a pentagon.
(From *Geodesics* by Edward Popko, © 1968, University of Detroit Press; reprinted
by permission of the publishers.)

War II by the Army Corps of Engineers in order to
conserve steel.

As an office building, it is ludicrous by today's
standards because of the length of corridors (17.5
miles) and the extraordinarily long walking dis-

tances that can be encountered on the interior.
One desk can be a half mile from another on the
same floor!

A far more pleasing shape is the plan of the
geodesic dome (Figure 5.8). The pieces are inter-

locking hexagons, but the shape is a pentagon.

THE PENTAGONAL RING

Equilateral triangles, squares, and hexagons fit together on a flat plane to make a continuous pattern covering all of the surface. Pentagons will not fit together on a flat plane with one exception: as shown in Figure 5.9, ten pentagons make a perfect ring. The inner shape is a decagon. The construction is one of many in which one polygon has the same edge length but twice the number of sides as the other.

This ring of ten pentagons will fit around any solid shape with a ten-sided equator: an icosa, icosadodeca, dodeca, pentagonal antiprism, decagonal prism, or decagonal pyramid. Figure 5.10 shows a pavilion with ten pentagonal pyramids assembled around the equator of an icosahedron. The whole pavilion with one major and ten minor galleries is assembled with modular equilateral panels. It is a pavilion fit for a king.

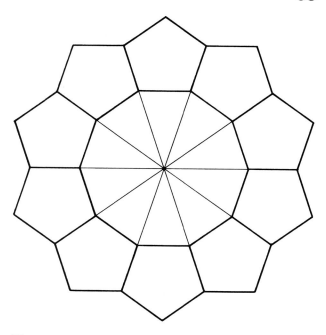

Figure 5.9
A ring of ten pentagons. The interior space is a decagon.

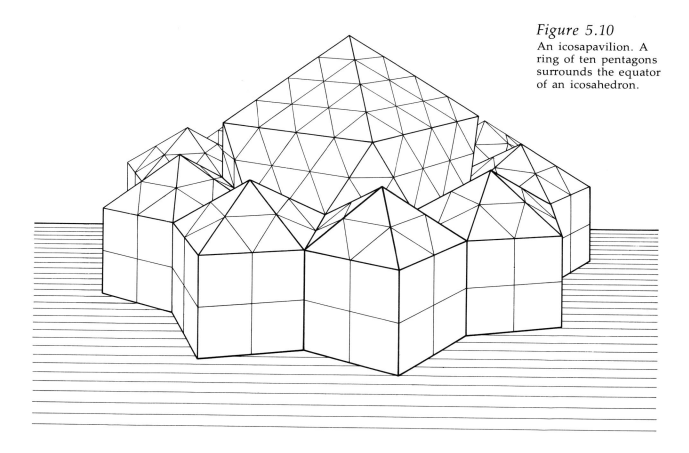

Figure 5.10
An icosapavilion. A ring of ten pentagons surrounds the equator of an icosahedron.

OTHER FIGURES OF FIVE

Figure 5.11, the "icosatwist," is an example of the rotation of squares within a square. In this case, five rotations bring the inner square into alignment with the outer square. Ten similar compositions can be made with angular changes of 2° to 45°.

The tile pattern of Figure 5.12 has the appearance of interlocking hexagons but consists of identical equal sided (but not equal angular) pentagons. The hexagonal patterns cross at right angles and the whole pattern can be fit into a square or subdivided into modular squares.

This unusual pattern, which is seen in street tiling in Cairo and occasionally in the mosaics of Moorish buildings, combines elements of four-, five-, and six-sided polygons and is another of the shapes in geometry in which, mathematically, the square root of seven plays a part. (The diagonal dimension of the rectangle enclosing the equilateral triangle includes the square root of seven.)

Like all other tile patterns, this pattern can provide the basis for a three-dimensional assembly of prisms. The photograph of a model (Figure 5.13) shows such an assembly with the pentagonal prisms proportioned to minimize the total surface area of each piece. With this proportion, the height as well as the dimensions of the base embodies the square root of seven.

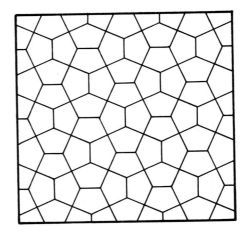

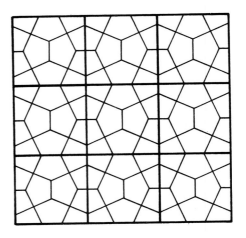

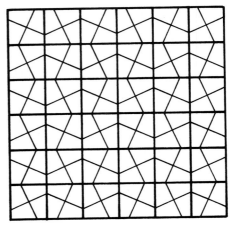

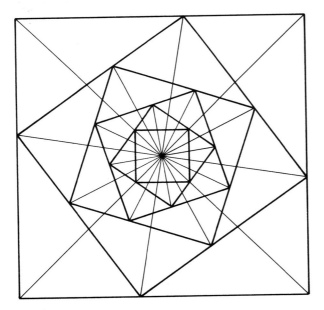

Figure 5.11
The icosatwist, an example of the rotation of squares within a square.

Figure 5.12
A pentagonal tile pattern with pieces equilateral but not equiangular. The pattern retains a right-angular relationship to the walls of a rectangular room and can be subdivided in several ways.

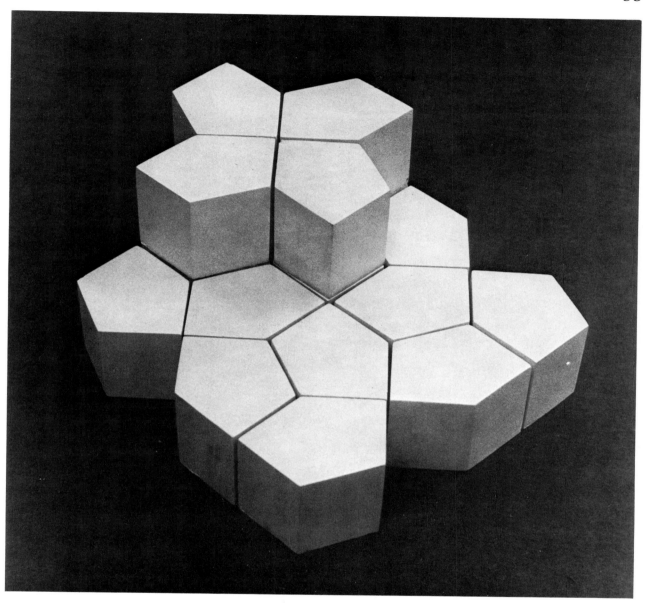

Figure 5.13
Prisms developed from the pentagonal tile pattern. (From *Space Enclosure Systems*
by Donald G. Wood, Engineering Experiment Station at The Ohio State
University.)

CHAPTER 6

THE 3–4–5 AND OTHER RIGHT TRIANGLES

Figures 6.1 and 6.2
Dwellings with 3–4–5
roof slopes by the
architects Sep Ruf and
Franz Ruf. (From *Das
Eigene Heim* by Guido
Harbers, © 1951 by Otto
Maier Verlag,
Ravensburg, Federal
Republic of Germany.)

When the "great encyclopedia of architectural geometry" is written, the simple right triangle, the favorite shape of Pythagoras, will be described as one of the two or three most important shapes in the history of constructions. The corners of our rooms and buildings are square because of it; bracing for dwellings, bridges, and coliseums depends on it; all of the regular and irregular polygons, and hence all of the classical solids, and all of the prisms, antiprisms, and pyramids depend on the mathematics and geometry of the right triangle. Even today, with the advent of drafting machines and computer plotting, an architect finds the elementary right triangle an extremely useful, if not indispensable, instrument of the profession.

The most common right triangles are the 45°, the diagonal half of a square, and the 30°–60°, the right triangle of both the equilateral triangle and the hexagon. These were used in the constructions of the 3/6 and 4/8 series of Chapters 3 and 4. There are, as well, the right triangles of the other regular polygons, but these are seldom found apart from the polygons.

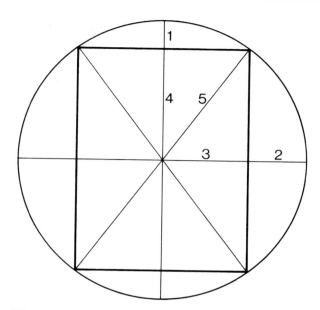

Figure 6.3
The 3–4–5 rectangle in a circle.

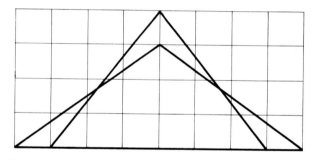

Figure 6.4
The two isosceles triangles made up of 3–4–5 right triangles. These triangles often are used as roof slopes.

THE 3–4–5 RIGHT TRIANGLE

Because of its mathematical simplicity, the right triangle with sides of three and four and a hypotenuse of five has as much appeal today as it had for the Babylonians 3000 years ago. Figure 6.3 shows the 3–4–5 right triangle, the rectangle of the 3–4–5, and its relation to a circle.

In architecture, the 3–4–5 often is used as a roof slope (Figures 6.1, 6.2 and 6.4), and it is also just the right slope for a comfortable stairway (Figure 6.5). In fact, with a single curious exception, it is the maximum slope permitted by most building codes in the United States. The exception is that steeper stairways can be used in homes, where persons least able to negotiate them, elderly people, pregnant women, young children, and furniture movers, must use them. This is something of an anomaly in the safety codes since the steeper stairs must surely cause many needless accidents and injuries in homes.

Stairways (Figures 6.6 and 6.7) offer unlimited design opportunities utilizing the 3–4–5 and other geometries.

Beyond these practical applications, the 3–4–5 can be used to make patterns of unusual interest, often with an undulating quality to the surface. Some of these patterns are illustrated in Figures 6.8, 6.9, 6.10, and 6.11. A subdivision of the 3–4–5 is seen in Figure 6.12.

INFINITE SERIES

The 3–4–5 stands not alone but at the head of an infinite series of right triangles which have whole

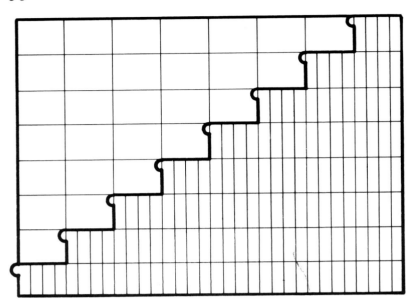

Figure 6.5
A stairway with a slope of the 3–4–5 triangle.

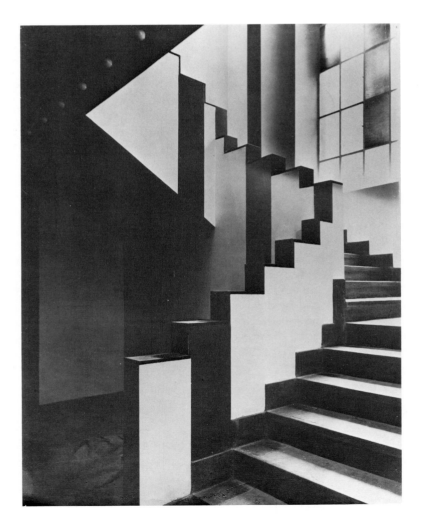

Figure 6.6
Stairway in the Café Aubette, Strasbourg, by Jean Arp and Sophie Taeuber-Arp, 1928. (Photo courtesy of Fondation Arp, Clamart, France © ADAGP, Paris, 1984.)

Figure 6.7
Stairway of the Laurentian Library by Michelangelo, 1558/59. (Photo by
Alinari/Editorial Photocolor Archives.)

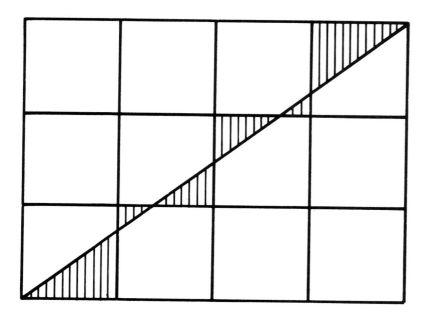

Figure 6.8
The diagonal subdivision of a 3 × 4
rectangle.

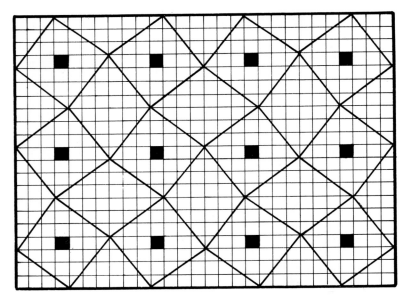

Figure 6.9
A quilt pattern based on the 3–4–5.

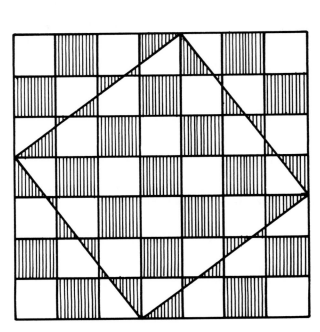

Figure 6.10
A 3–4–5 checkerboard.

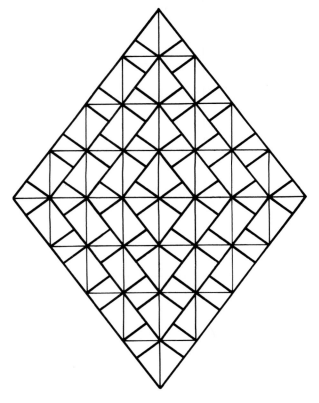

Figure 6.11
A 3–4–5 diamond design.

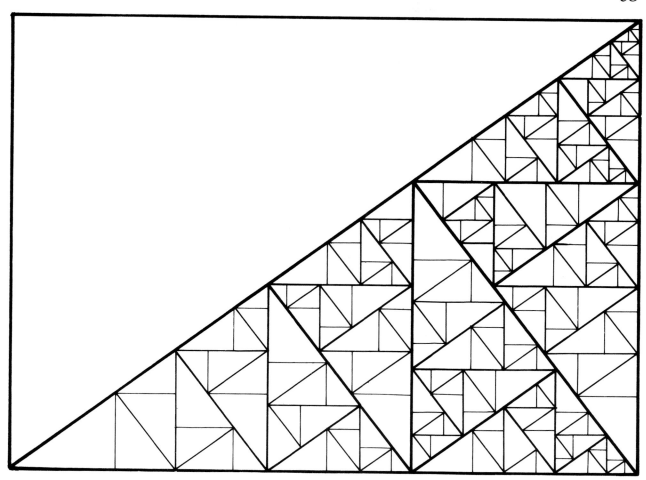

Figure 6.12

The subdivision of a 3–4–5 right triangle. In this illustration, a mechanical process of subdivision produces a delicate pattern of nonuniform line intensity. Any right triangle can be similarly subdivided.

numbers for sides (Figure 6.13). The short sides are 3, 5, 7, 9, . . . , the long sides are 4, 12, 24, 40, . . . , and the hypotenuses are 5, 13, 25, 41, ...

The fascinating aspect of this series is that while the slope is moving closer and closer to the vertical, the bottom leg is moving farther and farther from point zero, which means that the slope can never reach the vertical.

ILLUSTRATIONS OF THE PYTHAGOREAN THEOREM

The 3–4–5 is most familiar as the illustration of the famous Pythagorean theorem shown in clay tab-

lets dating from the end of the third millennium B.C. (Figure 6.14). This theorem states that in a right-angled triangle the square of the long side, the hypotenuse, is equal to the sum of the squares of the two shorter sides. The theorem usually is illustrated by squares, but equilateral triangles can be subdivided in the same manner as squares and can illustrate the theorem equally well (Figure 6.15). In fact, any polygon can be used since the theorem shows that the sum of two areas will equal a third when their sides are in a right-triangle relationship. In architectural terms, a right triangle can always be used to link three areas, two of which equal the third in size (Figure 6.16).

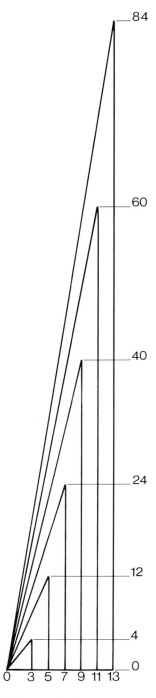

Figure 6.13
The infinite series of right triangles which have whole numbers for sides. The series begins with the 3–4–5.

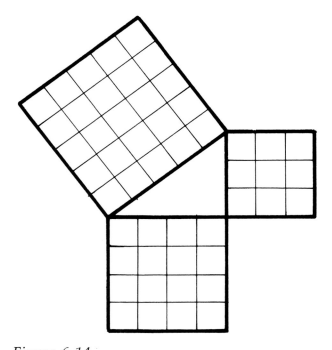

Figure 6.14
The conventional representation of the Pythagorean theorem.

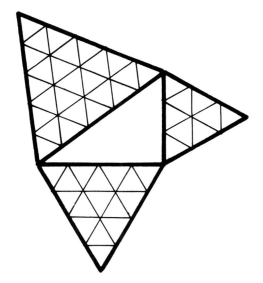

Figure 6.15
The Pythagorean theorem represented by the subdivision of equilateral triangles.

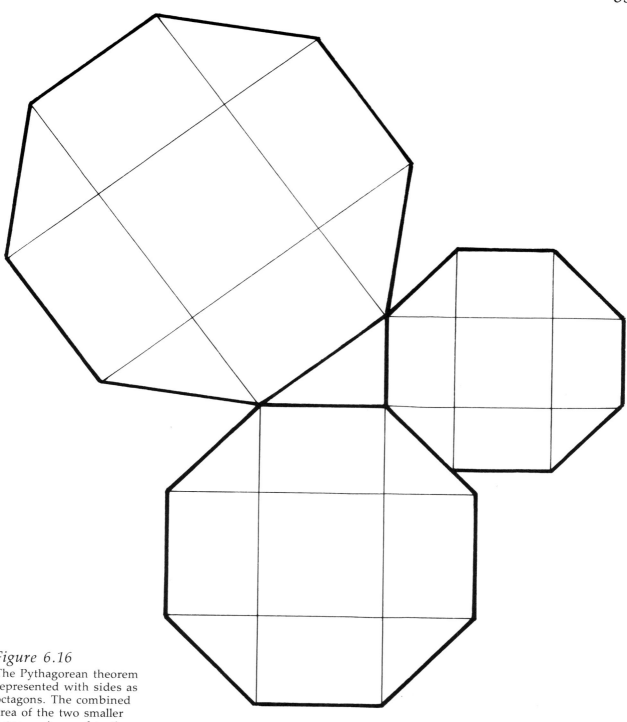

Figure 6.16
The Pythagorean theorem
represented with sides as
octagons. The combined
area of the two smaller
octagons is equal to the
area of the larger.

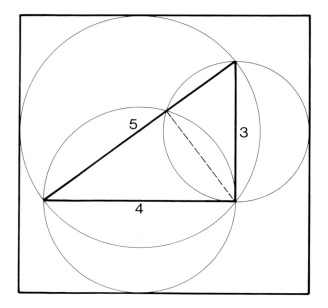

Figure 6.17
The 3–4–5 right triangle set in a square. With this construction, the side of the square is equal to 6.

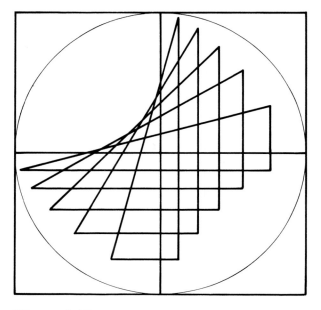

Figure 6.18
Right triangles set in a square. By this method of construction the perimeter of the square is always twice the perimeter of the right triangle set in the square.

3–4–5 AND 6

Going one step farther, the Pythagorean theorem can be illustrated by drawing circles with the sides of a right triangle as the diameters. This has been done with the 3–4–5 right triangle (Figure 6.17). Remarkably, the 3–4–5 fits perfectly in a square with sides tangent to the circles and, even more remarkably, it will be found that the side of the square will be 6, completing the 3–4–5–6 sequence often found in geometry. Any right triangle can be similarly "floated" in a square (Figure 6.18), and the perimeter of the square will always be twice the perimeter of the right triangle. A 0° right triangle is another instance of a two-sided line, shown both horizontally and vertically in this illustration.

RECTANGLE IN A SQUARE

The same geometric construction will "float" a rectangle of any proportion in a square space. Fig-

ure 6.19 is an illustration of a square surrounding a series of rectangles of different proportions. As with right triangles, the perimeter of the square is always twice the perimeter of the rectangle enclosed by this construction.

In these illustrations, geometric space around a triangle or rectangle is determined by the radii of circles. It is more difficult to determine the space needed around a building. Every building has a requirement for "breathing" space, a quantity of light and air around it and a certain distance between itself and its neighbors. This space may be needed on all sides or on only one or two sides, but it is an essential requirement for buildings intended for human use.

Particularly in the case of tall buildings, this breathing space is sometimes forgotten when buildings are planned side-by-side. Daylight and air are shut off from some parts of the building, and the street can become a canyon.

One of the two or three great advantages of tall

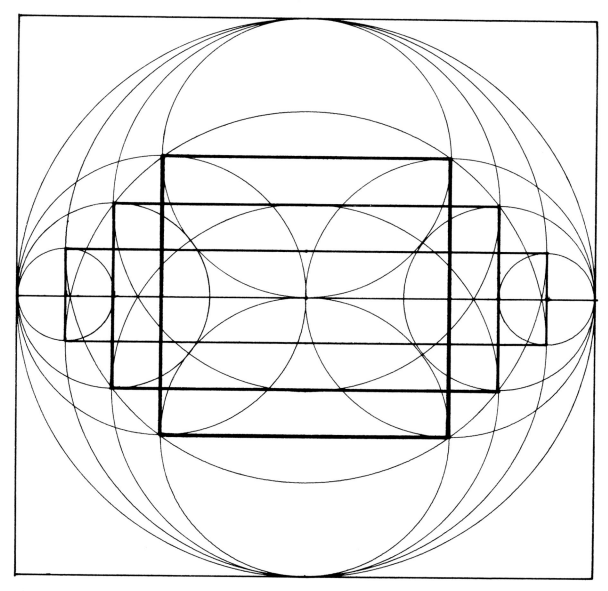

Figure 6.19
Rectangles set in a square. With sides as diameters of circles, any rectangle can be "floated" in a square. As with right triangles, the perimeter of the square is always twice the perimeter of the rectangle.

buildings is the amount of natural daylight and outside view afforded to their occupants, daylight and views which are cut off when buildings are jammed together. When the photograph of the Manufacturers Hanover Trust Company building in New York (Figure 6.20) was taken, an abundance of breathing space surrounded the building.

Figure 6.20
Former headquarters of Union Carbide Corporation, now owned by
Manufacturers Hanover Trust, by Skidmore, Owings and Merrill, 1960. Tall
buildings, like humans, need a certain amount of ''breathing'' space. (Reprinted
by permission from Union Carbide.)

CHAPTER 7

RECTANGLES

Figure 7.1
The Schroder House by Gerrit Rietveld, 1924. (From *The Work of G. Rietveld
Architect* by Theodore M. Brown. Copyright A. W. Bruna & Zoon,
Utrecht/Aartselaar.)

No SHAPE PLAYS AS IMPORTANT A PART in architecture as the rectangle. It is the most adaptable shape for human needs in all of geometry. Street grids, city blocks and lots, buildings, rooms, doors, windows, and furnishings are usually rectangular. Most building materials are rectangular. Most structural systems are rectangular. Within the rectilinear framework, almost everything needed for utility, comfort, and convenience can be integrated into a marvelously rational and harmonious system. Characteristically, the constructed environment is rectangular. It could be nothing else.

The geometry of the rectangle is elementary. It depends on a right angle. Size and the length-to-width ratio are the only variables. Size is determined by need, material limitations, site limitations, and budget. Proportion, the length-to-width ratio of a rectangle, is determined by similar factors but is governed as well by an aesthetic judgment as to fitness for the purpose. There is no one proportion of rectangle which satisfies all needs. There is a multitude of proportions, one of which may be better suited than the others for a particular purpose. The Gerrit Rietveld house of 1924 (Figure 7.1) is a veritable explosion of rectangles of different proportion in both the horizontal and vertical planes. The Chamberlain cottage by Marcel Breuer (Figure 7.2) is another composition emphasizing the rectangle as it occurs in the materials of construction.

IMPORTANT RECTANGLES

Beginning with the square, numerous rectangles can be identified as important in architecture. There are the rectangles of the regular polygons

Figure 7.2
The Chamberlain cottage in Weyland, Massachusetts, by Marcel Breuer, 1940. (Photo by Ezra Stoller, © ESTO.)

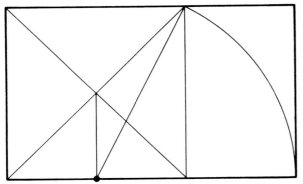

Figure 7.3
The rectangle of the golden mean. The length-to-width ratio is 1.1618:1. For much more about the golden mean as well as the aesthetic appreciation of mathematics, see *The Divine Proportion* by H.E. Huntley (Dover Publications, 1970.)

hexagon. The off-square rectangle surrounding the pentagon is also important. Rectangles that can be divided evenly into square modules comprise yet another series of value to an architect.

The most famous rectangle in architecture is the golden mean (Figure 7.3). It is fashionable these days to discount proportion per se as not having aesthetic validity in an age dominated by many other interests, but the fact remains that this particular rectangle has survived because it is pleasing to the eye, in the abstract, and has as well the mathematical intrigue of the number ϕ.

described in Chapter 2, of which the rectangle of the pentagon is a prime example. The slightly off-square rectangle encompassing the equilateral triangle (Chapter 3) is linked closely with the many patterns of the equilateral triangle and the

SUBDIVISION OF RECTANGLES

The subdivision of rectilinear space and rectilinear planes is a principal concern and challenge of architects. Pure proportion and nonsymmetrical balance as they occur in Mondrian's *Composition with Red, Blue, and Yellow* (Figure 7.4) is a goal of architecture, but the final judgment of the quality of an architectural space usually is based on suit-

Figure 7.4
Composition with Red, Blue, and Yellow by Piet Mondrian, 1930. (From the collection of Mr. and Mrs. Armand P. Bartos.)

ability for purpose rather than pure proportion in the abstract. Nonetheless, differences in visual balance and proportion are discernible by the eye and cannot be ignored in the design of rectangles for human use. The choice of proportion is often a fine one, but this choice will greatly influence the measure of satisfaction to be derived from the use of the space or the appearance of a plane.

Louis Sullivan in his design of the facade of the Carson Pirie Scott store in Chicago (Figure 7.5) emphasized broad window proportions in harmony with the horizontal girders and the space

Figure 7.5
The Schlesinger Meyer (now Carson Pirie Scott) Department Store in Chicago by Louis Sullivan, 1899–1904. (Photo by John Szarkowski from *The Idea of Louis Sullivan,* University of Minnesota Press, 1956.)

within. The repetitive composition has both a major and a minor theme in well-proportioned rectangles.

THE PROPORTION OF ROOMS

A box (Figure 7.6) has the obvious dimensions of length, width, and height, and it also has the ever-present fourth dimension of wall thickness. Adequately sized, properly proportioned, well-built, with skillfully placed furniture, doors, windows, with carefully selected floor, wall, and ceiling materials, with complementary furnishings, fixtures, and accessories, dry, and with comfortable temperature and humidity, a box is transformed into a room that suits a human purpose.

Such a room is the children's bedroom by Heinrich Wurm (Figure 7.7). Notice the recurrence of rectangular shapes in this small room.

The sixteenth-century Italian architect Andrea Palladio listed seven shapes he considered ideal for rooms (Figure 7.8). Of these, the three-by-five room with a proportion just slightly greater than that of the golden mean was his favorite, prob-ably because the three-by-five proportion yields an odd number of equal subdivisions on both the long and short sides, assuring that an opening and not a column will always occur in the middle of either wall. None of Palladio's room proportions is greater than 2:1.

The proportions of a room are twofold. The length-to-width ratio of the floor is one proportion, usually determined by the functional requirements of the room. The relationship of the height of the room to the length and width of the floor is another, sometimes independent, consideration. In terms of space enclosure, the optimum height for a given floor proportion is defined in Chapter 11.

RECTANGULAR BUILDING SHAPES

While room shapes are usually rectangular and seldom greater than 2:1 in proportion, building shapes, especially low buildings, cover virtually the whole range of rectangles, from perfect squares to the mile-long assembly plants of World War II. L, T, U, and similar shapes are equivalent

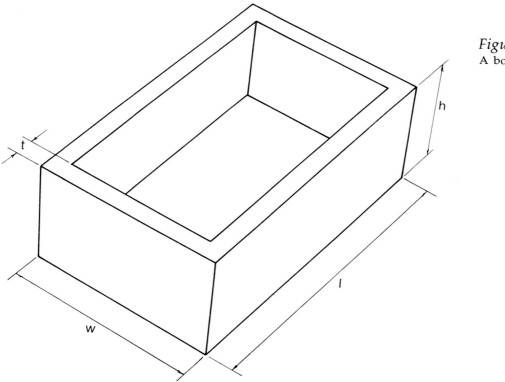

Figure 7.6
A box.

Figure 7.7
A children's bedroom in Ravensburg, West Germany, by Heinrich Wurm, Architect. (From *Das Eigene Heim* by Guido Harbers, © 1951 by Otto Maier Verlag, Ravensburg, Federal Republic of Germany.)

to long rectangles. Like long rectangles, they provide a lot of perimeter with respect to the floor area. Courtyard plans are especially advantageous in this respect.

The perimeter efficiency of various rectangular shapes is seen in Figure 7.9. This chart shows in detail the rectangles seen only in outline form in Chapter 2. Length of perimeter, the amount of outside wall or window space on the floor of a building, may be of great importance to an architect or none at all, depending on the function of a building.

A rectangular floor or building with the smallest amount of perimeter surface is the most economical to construct, maintain, heat, and cool. At the same time, the building with the least perimeter will provide the least amount of daylight natural ventilation, and outside views for its occupants, factors important to the quality of interior spaces.

Warehouses, which have no special requirement for daylight, typically are box-shaped struc-

tures with minimum perimeter. On the other hand, hotels, hospitals, houses, and apartment buildings need a lot of perimeter to provide the occupants with sunlight, air, and views.

Airport terminals seem to have the greatest requirement for perimeter length. This is not for the benefit of passengers but because parking today's airliners, with their huge wingspan, requires an extraordinary amount of curb space. Since traditional airport design calls for aircraft to dock alongside the terminal, the predictable result is extremely long walking distances for passengers, which can be overcome only by the use of a conveyance.

In the design of office buildings, there are two opposing viewpoints on perimeter. Some say that daylight and view are relatively unimportant for office workers, that a properly "landscaped" interior may be substituted, and therefore economical bulky shapes are best. Others maintain that office buildings should indeed provide occupants with

T H E moſt beautiful and proportionable manners of rooms, and which ſucceed beſt, are ſeven, becauſe they are either made round (tho' but ſeldom) or ſquare, or their length will be the diagonal line of the ſquare, or of a ſquare and a third, or of one ſquare and a half, or of one ſquare and two thirds, or of two ſquares.

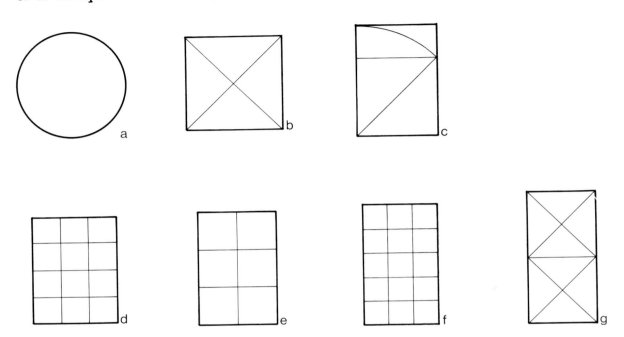

Figure 7.8
The room proportions of Andrea Palladio (1518–1580), as described in his book
The Four Books of Architecture. (Dover Edition © 1965.)

as much daylight and view as possible. The deciding factor in designing buildings today may be the need for energy conservation. In that case, the greater the perimeter, the greater the opportunity for passive solar heat, daylight, and natural ventilation, provided that effective sun control is part of the architectural design.

A square has the least perimeter of any rectangle and can be subdivided into square modules of any size. Thereafter, perimeter length can be increased in two ways. Either the proportion of the rectangle can be increased or the area can be divided into a number of smaller areas, as illustrated by the groupings of two, three, and four squares at the lower left edge of the chart in Figure 7.9.

The greater the length of perimeter, the greater will be the number of choices in shape available. Once perimeter requirements have been determined, the shape of a long rectangle may be changed to an L, T, U, or other linear configuration without affecting the length of the perimeter.

On any horizontal line of the chart the length of perimeter is the same but there are other geometric differences between the shapes. These are differences in compactness, walking distance, corridor length, and number of outside corners available, which will influence the final choice of shape.

The groupings of two, three, and four squares at the lower left edge of the chart, as explained, show the huge increase in perimeter that occurs when a large area is divided into smaller separate blocks. They illustrate an important point: the huge increase in perimeter that occurs in multi-story buildings. As far as perimeter is concerned, it makes no difference whether the separate blocks are on the ground or are floors of a building stacked one on top of another. The four squares at the lower left corner, representing a four-story

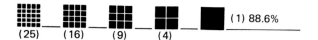

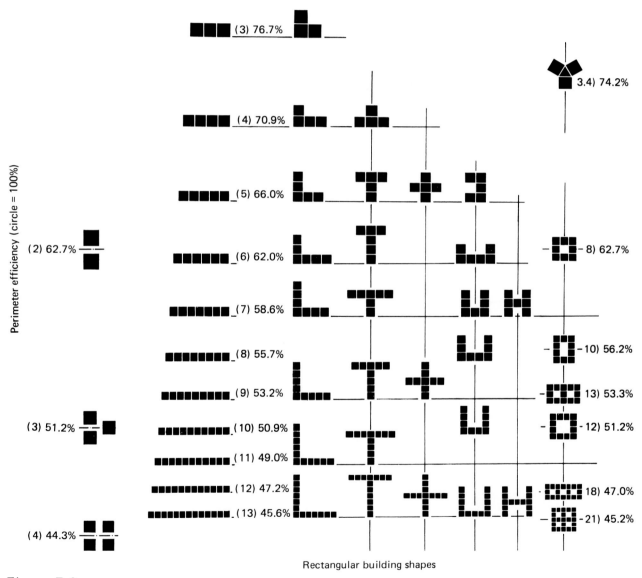

Perimeter efficiency (circle = 100%)

Rectangular building shapes

Figure 7.9

The perimeter efficiency of rectangular building shapes. The universe of
rectangular shapes has been simplified on this chart by selecting only those that
can be divided evenly into square modules.

building, have more perimeter (daylight) than any of the long rectangle, L, T, U, and other one-story shapes shown on the bottom line.

Tall buildings will always provide more daylight and openness than low buildings of the same total floor area. For instance, the eighteen glass-enclosed floors of the Lever House tower (Figure 7.10) have a length-to-width ratio of 3:1. A one-story building with the same floor area *and* the same amount of window area would have a length-to-width ratio of nearly 100:1! Zoning restrictions often arbitrarily limit the height of buildings, but the functional advantages of height are compelling. Tall buildings reduce vast floor areas into small, manageable spaces; they provide the maximum potential for daylight, natural ventilation, and views; they greatly reduce walking distance with reliance on energy-efficient electric

Figure 7.10
Lever House in New York by Skidmore, Owings and Merrill, 1952. (Photo by Ezra Stoller, © ESTO.)

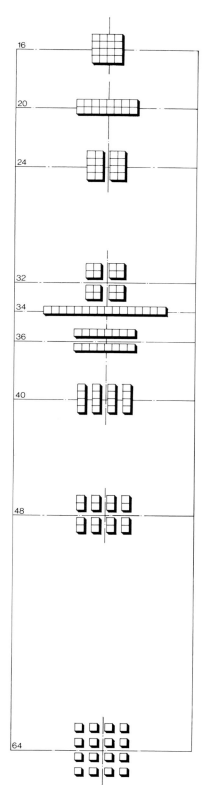

Figure 7.11

Nine arrangements of sixteen squares. The length of exposed perimeter changes from the least at the top to the most at the bottom of the chart. The shapes represent arrangements of separate floors of buildings or of separate dwellings.

elevators; and they minimize foundation and roof areas—important considerations even in the desert.

Provided that adequate light and breathing space are provided around each structure, tall buildings offer a rational and humane solution to the problem of providing comfortable working environments in a compact area to meet the needs of an expanding population.

THE ILLUSTRATION OF SIXTEEN SQUARES

Another way to appreciate the importance of perimeter is to look at the possible arrangements of sixteen square blocks as shown in Figure 7.11. The squares can represent blocks of office space or they can represent individual dwellings.

At one end of the scale, the squares are separate blocks representing either sixteen separate square buildings or a sixteen-story square tower. In either case, there are sixty-four corners and sixty-four units of exposed perimeter. At the other end of the scale, the sixteen squares are arranged in their most compact form, a single large square representing a one-story building with only four corners and sixteen units of perimeter. The floor area is the same but one arrangement has sixteen times the number of corners and four times the amount of perimeter as the other. Between these two extremes are various arrangements with differing amounts of perimeter, different numbers of corners, and important geometric differences in the walking distance required to go from one place to another.

As dwellings, the sixteen separate squares at the lower end of the scale represent detached houses. The next grouping in pairs is an approximation of duplex arrangements. Then there are row arrangements and the New England "quads" or four-plexes. All of these have advantages and disadvantages, but by far the most livable and urbane, especially in a time of land and energy conservation, is still the row house (Figure 7.12). The extreme at the upper end of the scale is a "beehive" in which some of the squares representing dwellings have no outside exposure, an untenable arrangement for human habitation.

As blocks of office space, the sixteen squares

representing a sixteen-story office building at the bottom of the chart have the greatest amount of perimeter (light, air, and view), the greatest number of corner offices, use the least amount of land when stacked one on top of another, and have the shortest walking distance between corners when linked by elevators. The compact square at the upper end represents the "landscaped interior" or warehouse concept of accommodating human activities, has the least perimeter, fewest corner offices, uses the most land, and has the greatest walking distance between diagonally opposite corners of one floor.

Figure 7.12
Baltimore rowhouses. "[The 18th century] invented the first really comfortable human habitations ever seen on earth, and filled them with charming fittings. . . . The 18th century dwelling house has countless rivals today, but it is far superior to any of them as the music of Mozart is superior to Broadway jazz. It is not only, with its red brick and white trim, a pattern of simple beauty; it is also durable, relatively inexpensive, and pleasant to live in. No other sort of house better meets the exigencies of housekeeping, and none other absorbs modern conveniences more naturally and gracefully. Why should a man of today abandon it for a house of harsh masses, hideous outlines and bald metallic surfaces? . . . I can find no reason in either faith or morals. The 18th century house fits a civilized man almost perfectly. He is completely at ease in it. In every detail it accords with his ideas."—H. L. Mencken. (From *Those Old Placid Rows* by Natalie W. Shivers, Maclay & Associates, 1981. Photo by William L. Klender, Baltimore Sunpapers.)

CHAPTER 8

THE DIAMOND AND THE DIAGONAL

Figure 8.1
An airview of a typical rectangular pattern of streets and city blocks. The
rectangular grid may be prosaic but it is extraordinarily useful. (Photo by Pacific
Aerial Surveys, Oakland, California.)

THE RIGHT-ANGLED RECTANGULAR STREET GRID (Figure 8.1) has endured for centuries as the framework for urbanization and civilization. Just in the past fifty years has it been greviously disrupted in a few places by the wide-sweeping curves of high-speed freeways, massive structures which are a boon to drivers and commerce but an assault on the traditional ground-level marketplace. In spite of this, there is little doubt that the grid will survive; it is too rational.

The grid is simple, direct, readily comprehensible, easy to address, efficient in land use, adaptable to a multitude of uses, can be started and stopped at any point or extended in any direction, and can overlay almost any shape. Alternative routes to any point are available, so that if one becomes blocked another is open. It is both mundane and indispensable. The Austrian architect, teacher, and city planner (before there were city planners) Camillo Sitte abhorred the grid. The famous Swiss-French architect Le Corbusier loved it. The grid remains an extraordinarily useful adjunct of urban life.

The width of streets and sidewalks, the size and proportion of blocks, the orientation of the grid with respect to sun and wind, are matters to be considered in developing a new grid. In an existing city these things have been decided. There are, however, two geometric principles related to travel across a grid which are valuable in planning a new grid or in planning on an existing grid. These are the principles of the diamond and the diagonal.

THE DIAMOND

As urban dwellers, most of us live near our place of employment, shopping center, school, recreation area, and so on. There are generally accepted rules about how near these places should be for optimum utilization. Bus stops, convenience shopping, elementary school, park, and other neighborhood places should be within an easy walking distance of the doorstep, usually taken as five minutes or so at an assumed walking rate of 4 feet per second. A junior high school, major recreation area, and automobile service station should

be within a twenty-minute walk and easy bicycle range. People living closely together in wood frame houses, as they do in San Francisco, are safer when there is a firehouse within, say, three minutes at 25 miles per hour. Ideally, the place of work ought to be within a thirty-minute commute. Once formulated, these rules determine the limits of the area around a place, the tributary area, where people can live and be within a reasonable travel time or distance based on the means of transportation.

If one could travel from place to place by helicopter, that is, in a straight line, the shape of the area within a given distance to a place would be a circle. Conventional travel by foot, bus, or car is down on the streets and sidewalks, however, and the travel route will usually be on a grid pattern. The shape of the tributary area with respect to a place will not be a circle. The walking or driving distance from one place to another on a grid is the sum of the lengths and widths of the blocks traversed and not the straight line distance from point to point. It is the length of the path in the first or second illustration in Figure 8.2 but not the path shown in the third. Consequently, the locus of points equidistant from a point on the grid is a square turned at 45° to the street grid, that is, a diamond shape (Figure 8.3), not a circle. As Figure 8.3 shows and calculations will confirm, a circle includes more than half again as much area as actually lies within the walking or driving distance indicated by its radius.

Figure 8.4 shows diamond-shaped zones of equal walking distance to and from the center of a rapid transit platform in a neighborhood. The density of development is graduated from highest in the inner zone to lowest in the outer zones. The result is a three-dimensional clustering effect with intense activity at the center but gradually diminishing activity in the outer zones. The black edges represent the location of shops and stores.

In this plan, walking distance is used as a basis for establishing zones of differing density around an important amenity in a way that intensifies the use of the amenity. As density diminishes, the height limit also steps down from the center.

The open space at the center is a key element in the success of the plan.

A central square with a statue or fountain near

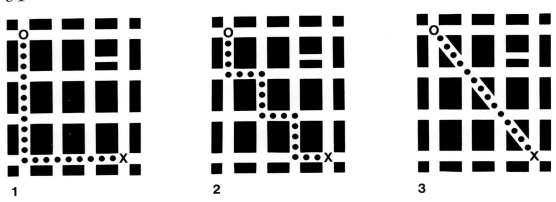

Figure 8.2
The pedestrian path on a rectangular grid. The walking distance between O and X is the length of the path shown in the first or second diagram but not the path shown in the third.

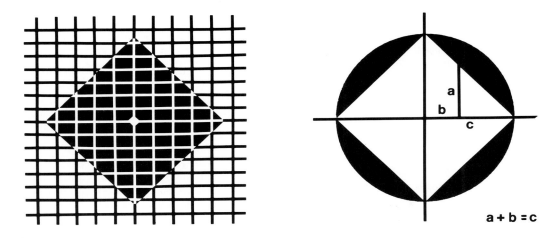

$$a + b = c$$

Figure 8.3
The diamond shape of the area within a given travel distance to the center. With this shape, $a + b = c$.

its center and with important buildings surrounding it (Figure 8.5) is still one of the most practical and pleasant ways to structure a town, campus, or shopping center. It is compact, minimizes walking distances, offers great visibility, can accommodate large numbers of people, and is susceptible to immense variety in design. Most important, the central space provides an opportunity for social focus and community cohesion through visual contact and awareness.

CAMPUS GEOMETRY

An application of the diamond theory on a larger scale is in campus planning where time and pedestrian movement have traditionally been given special emphasis. Numerous campus plans place the library at the center of a hypothetical circle encompassing the other academic facilities. The diameter of the circle is the distance that can be walked in ten minutes, the time alloted between scheduled classes. No one is more than five minutes from the library. Thus defined, the size of the central campus is 104 acres.

In fact, unless the campus has an infinite system of walks uninterrupted by the buildings themselves, the plan has only symbolic appeal. With the usual rectangular system of walks, the distance from one building to another within the circle can

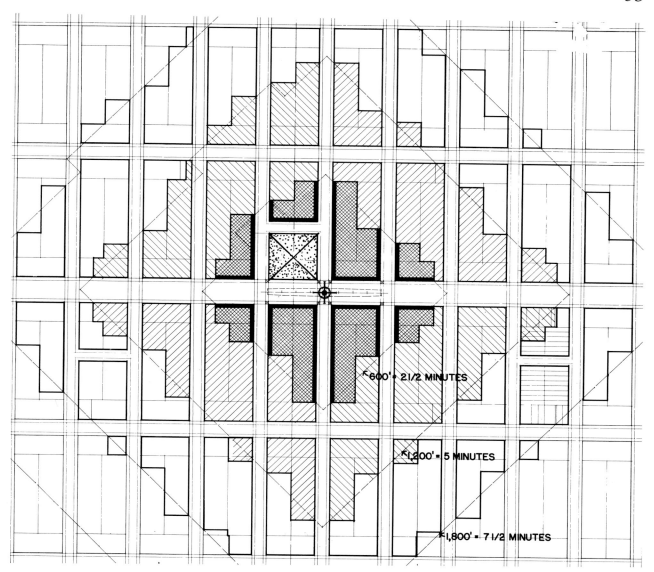

Figure 8.4
The diamond shapes of concentric zoning with respect to a rapid transit station
based literally on walking distance and existing property lines.

be fourteen minutes; thus the students will be late for class and the library at the center will be an obstruction in everyone's path.

Assuming at the outset a rectangular system of walkways and the same ten-minute walking time, the shape of the central campus area will be a 66-acre diamond. Buildings within that diamond will, in fact, be not more than ten minutes apart. A major open space is at the center with the library and other important buildings arranged around it and not at it. Again, the circle introduces a 40%

error, not even a close approximation, whereas the diamond concept results in a more compact campus with a significant conceptual shift. Other kinds of urban complexes follow this same principle of design.

A CITY PLAN CONCEPT

With the tributary area accurately defined as a diamond, the uniform distribution on a grid of

Figure 8.5
An example of a city square. (Engraving from Fernando Leopoldo, *Firenze Citta Nobilissima*, Florence, 1684, courtesy of the Fine Arts Library of the University of Pennsylvania.)

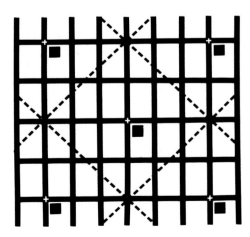

Figure 8.6
The staggered pattern of places uniformly distributed with respect to travel distance on a rectangular grid.

similar places such as neighborhood centers or firehouses produces the staggered pattern of Figure 8.6. The broken line on the grid defines the tributary area, but it has no reality itself. That is, it is not a road, walk, fence, or anything that can be seen. It may be a ridgeline if the tributary area is also a drainage area with gravity flow to the center.

With this staggered pattern of distribution, the geometry of the diamond can be used to construct models for new cities. Figure 8.7 shows a city with twenty-four residential neighborhoods on both sides of a central business area. Major streets are spaced at 1200-foot intervals each way—five minutes apart on foot.

The vital part of the town center (crosshatched area) is consolidated around a central open space. Every place within the crosshatched area is within

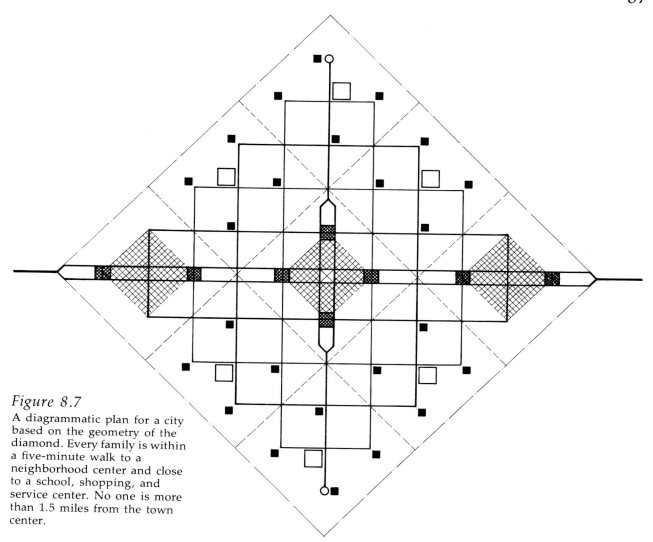

Figure 8.7
A diagrammatic plan for a city based on the geometry of the diamond. Every family is within a five-minute walk to a neighborhood center and close to a school, shopping, and service center. No one is more than 1.5 miles from the town center.

a five-minute walk to the central square. Places within the remainder of the town center have a maximum ten-minute walk to the square and are no more than twenty minutes on foot from any other place in the center. Parking garages are provided in the center of each quadrant so they also are no more than five minutes from any place.

The town center is flanked by two other centers identical in size. These centers may be industrial, medical, governmental, institutional, or recreational but they follow the same rules for compactness as the central business area with limits based on walking distance.

The twenty-four neighborhoods have a park and other neighborhood facilities within an easy walking distance of every dwelling. In turn, four neighborhoods form a large precinct or district with major shopping and schools within a ten-minute walking distance and an easy drive. Only the major through streets are shown in the diagram, not the substreets of the neighborhoods. No one in a neighborhood is more than a mile and a half from the downtown center by bus or car.

It is improbable that a real city would have a pattern of such regularity, neglecting as it does a multitude of other considerations. But the plan

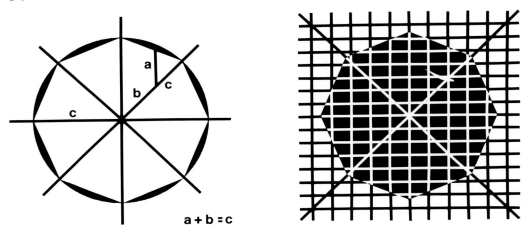

Figure 8.8
The octagonal geometry of the diagonal on a grid. In this case also, $a + b = c$.

Figure 8.9
The four diagonals of the 1821 plan for Indianapolis, Indiana. Diagonals need not cross at the center to be fully effective. They need only lead to and from an area. (Map provided by Historic Urban Plans, Ithaca, New York, from the original in the Indiana State Library.)

does illustrate the uniform distribution of places on a rectangular grid with three different travel–time relationships. Major emphasis is on pedestrian, and not automotive, movement. It is a modular plan based on the first and most important city planning constant, that of walking distance. The end result is a compact city with maximum land and energy conservation.

THE DIAGONAL

Diagonal streets supermposed at 45° on a rectangular grid change the geometric shape of the tributary area from a diamond to an octagon (Figure 8.8), a shape that approaches the characteristic compactness of a circle. The area of the octagonal shape created by the diagonals will be 40% larger than the area of the diamond shape without the diagonals. Thus in a city, neighborhood, or campus, the area within a given distance to the center can be increased 40% in any one quadrant or the city as a whole by the addition of a diagonal street.

The four diagonals that were introduced in the 1821 plan of Indianapolis (Figure 8.9) serve exactly the intended purpose. Diagonals need not cross over an area to be fully effective; they need only lead to and from an area.

The effectiveness of a diagonal does depend on access at frequent intervals along its entire length. A diagonal having grade intersections in the outlying areas, but changing to an expressway with limited access over the more congested area, will not achieve the intended results.

Columbus Avenue, a diagonal street in San Francisco (Figure 8.10), was reputedly constructed

Figure 8.10

An airview of Columbus Avenue in San Francisco. (Photo by Pacific Aerial Surveys, Oakland, California.)

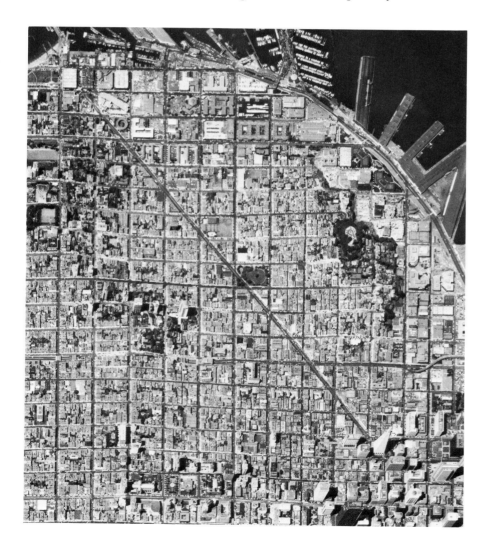

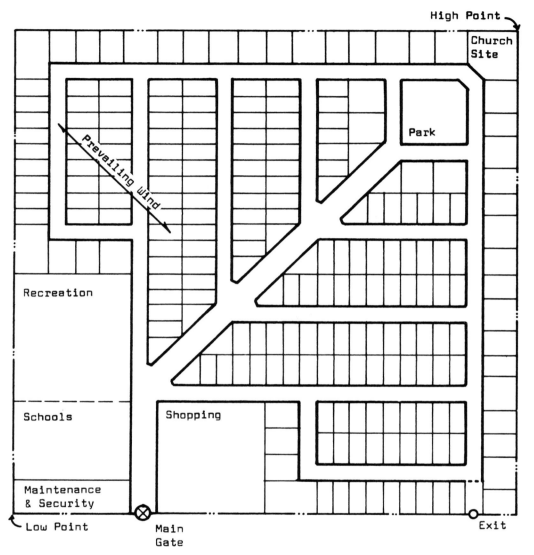

Figure 8.11
A subdivision plan
with one diagonal.

to accommodate an influential developer whose project was failing because it was too remote from the heart of the city. The avenue reduces walking and driving distances for thousands daily.

On a smaller scale, the subdivision plan (Figure 8.11) illustrates the effective use of a single diagonal to shorten travel distance to the community center—school, shops, and recreation center. Compared with other plans for the same site, this plan has shorter travel distance and fewer turns to each lot (on foot or by car), excellent traffic distribution, and appealing pedestrian flow to the center. The geometry is similar to that of the maple leaf (Figure 8.12).

DIAGONALS AND THE RADIAL CITY

A single diagonal street in each quadrant is worth consideration because of the magnitude of the change, a 41% increase in area, which is more than enough to compensate for some disruption in the grid pattern and for a small loss in area used by the street itself.

Two diagonals 30° apart will increase the tributary area by 44%. But this is only 3% more than can be accomplished by one diagonal. This 3% gain may not outweigh the loss in area and disruptive effect of a second diagonal.

The addition of more diagonals brings progres-

Figure 8.12
The 45° diagonal geometry of a maple leaf.

sively less gain and more disruption and loss of land. The limit is a near perfect circle with a multitude of diagonals, the radial plan. While the tributary area of a radial plan is 57% greater than the tributary area of a rectangular grid without diagonals, it is only 15% greater than the gain from a single diagonal. This 15% gain must be weighed against an enormous loss in area (nearly 100% at the center), the awkward size and shape

of blocks, and the difficulty encountered in moving in a direction other than to the center. With diagonals there is a point of diminishing return, and it occurs after the first one.

The ground plans for Urubupunga in Brazil (Figure 8.13) and Nahalal in Israel (Figure 8.14) are examples of radial plans with a multitude of diagonals.

Figure 8.13
Urubupunga, a temporary town with a radial plan constructed to house workers during the construction of a hydroelectric complex. The city no longer exists. (Photo courtesy Companhia Energetica de Sao Paulo, Brazil.)

Figure 8.14
Nahalal in Israel planned in the 1920s with a radial concept. (From *The Matrix of Man* by Sibyl Moholy-Nagy © 1968 Frederick Praeger.)

CIRCLE, SEMICIRCLE, AND OTHER CURVES

THE ART OF ARCHITECTURE, because of the nature of the materials and processes, is primarily the art of straight lines. Where there is a will, however, there is a way and architects throughout history have found ways to introduce curves into our environment. Cities as old as Baghdad began as circles (Figure 9.1); the great stone monuments, arches, and amphitheaters of antiquity are circles and semicircles; in modern times the architect Le Corbusier, among others, mixed curved elements into a rectangular framework almost routinely.

Curves, of course, may be required for functional and aesthetic reasons. Carousels and ferris wheels must be circular. Dome structures dictated by the necessity to span huge spaces are round. People and vehicles turn on curves, not at right angles, and these turns may determine the walls of architecture.

There are many kinds of curves: circles and semicircles, ellipses and semiellipses, catenaries, hyperbolas and parabolas (although not many of these), spirals, and an array of irregular or free-form curves, all with special features. Some of these curves, especially those in which the radius of curvature is constantly changing, are difficult to construct. Nonetheless, they are all called simple curves because the curvature is in one plane only. Spheres, hemispheres, domes, hyperbolic paraboloids, complex catenary structures, air-supported structures, and similar shapes which curve in more than one plane are compound curves and a part of solid geometry.

THE CIRCLE

The circle (Figure 9.2) is the simplest of the two-dimensional shapes and the easiest to draw on paper or inscribe on the ground. One can imagine the pleasure the ancients had in laying out the circles of Stonehenge (Figure 9.3). They needed only flat ground, a length of rope attached to a

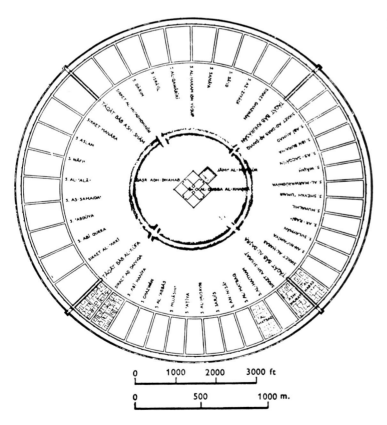

Figure 9.1
The plan of Baghdad.

stake, someone to walk it around keeping it always taut, and someone to mark the path. Arranging the stones was a more difficult task! Circles on paper are always a pleasure to draw. The pencil line is everywhere the same distance from the center and returns unerringly to its starting point.

The circle encloses a given area with the least perimeter or circumference and is the most compact of the plane geometric shapes. A circle has only one dimension, its radius or diameter, and it is located by only one point, its center. A complete circle appears to have no beginning and no end,

and it has no corners. It is nondirectional unless an axis is drawn or otherwise indicated in the plane of the circle. King Arthur's position at the round table (Figure 9.4) is marked only by a special chair. The round table itself has neither head nor foot.

Circles do not have an axis in the ground plane but may have a vertical axis at the center, either real, as in the case of a wheel, or imagined, as in a circular room where the walls everywhere focus to the center. For maximum effect, an architect will often concentrate visual attention on this imaginary vertical axis.

From the outside, the walls of a circular building face equally in every direction and are powerful diffusing, not focusing, elements.

Geometrically, a circle is an ellipse in which the two focal points coincide at the center. It is the fullest and the most compact of the ellipses. Its complement is an extremely flat ellipse (a straight line) in which the focal points are as far apart as possible.

A circle may also be a regular polygon with a great many sides. In fact, beginning with the pentagon, regular polygons represent successive approximations to the circle. Since it is often easier to build the straight sides of a regular polygon than the continuously curving sides of a true circle, these approximations (shown in Chapter 2) are useful. A twelve-sided polygon (dodecagon) is a 98.8% approximation to the characteristic compactness of a circle, and a forty-sided polygon is a

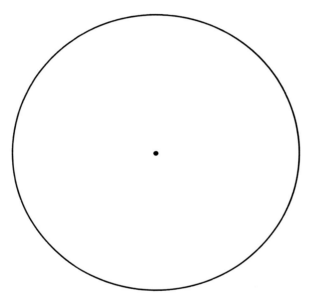

Figure 9.2
The circle.

Figure 9.3
Stonehenge, a circular monument of stones built on the Salisbury Plain of England about 1800–1400 B.C. The diameter of the circle is 97 feet. Although they appear gargantuan, the height of the stones is only 13½ feet. (Photo British Crown Copyright, reproduced with permission of the Controller of Her Britannic Majesty's Stationery Office.)

WHEN THE CALL FOR A MEETING OF THE ROUND
TABLE IS SENT FORTH, ONE HUNDRED OF THE
BRAVEST KNIGHTS IN ALL THE LAND ATTEND.

Figure 9.4
King Arthur's Round Table. The main point about a circular table is that
everyone can see and be seen. It is the perfect shape for an assembly of equals.
(From *Prince Valiant* by Hal Foster, © 1979 King Features Syndicate, Inc.)

99.9% approximation, close enough for all but the most discerning eye.

When uniform pressure is exerted on a circle, either from the inside or from the outside, the circle is the strongest of shapes. For this reason, it is the shape of oil tanks, gun barrels, and shafts.

For the same reason, masonry arches, possibly the most eloquent shapes of architecture, are semicircular.

The square and rectangle can be geometrically subdivided in a number of ways but there are only one or two ways to subdivide the circle and el-

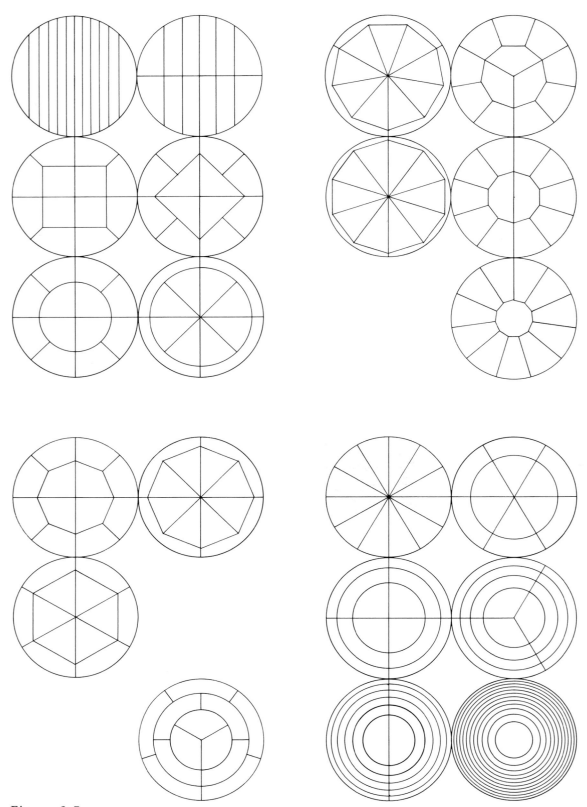

Figure 9.5
Some of the more than 600 ways to divide a circle into twelve equal parts with a straightedge and compass.

Figure 9.6
Model of the Oakland–Alameda County Coliseum by Skidmore, Owings and Merrill, 1967. The roof of the arena has been removed to show the interior. (Photo by Dwain Faubion.)

lipse, which limits the use of these shapes. There is at least one remarkable exception, however. A circle can be geometrically subdivided into twelve equal parts in more than 600 ways (Figure 9.5). With respect to this one number of equal subdivisions, a circle has extraordinary versatility.

The equator of the earth is a circle; the moon and the sun with their halos are apparent circles; a pebble dropped into a still pond generates circles; there are other circles to be found in nature. In architecture, besides monuments, round tables, and cities, circles have been used in the plans of stadiums and arenas (Oakland Coliseum, Figure 9.6), elegant cast iron tree grates (Paris, Figure 9.7), inviting stairways (Palladio, Figure 9.8),

Figure 9.7
A Parisian cast iron tree grating. (From *Cities* by Lawrence Halprin © 1963, Reinhold Publishing, reprinted by permission of Van Nostrand Reinhold Company. Photo by Lawrence Halprin.)

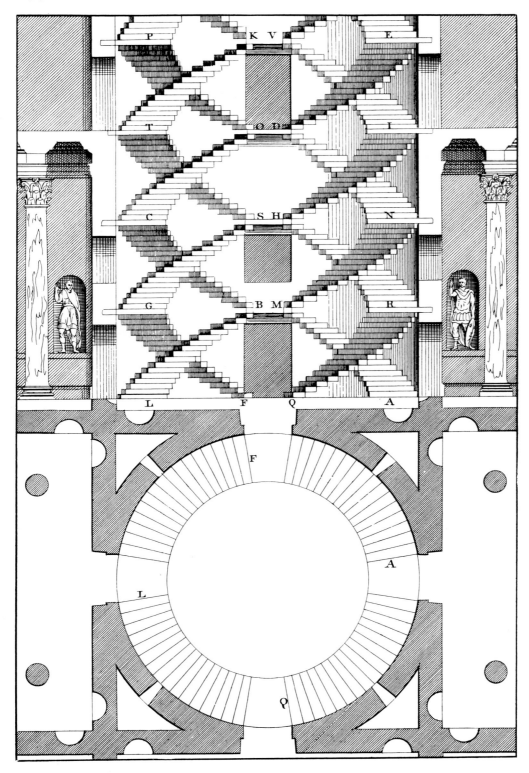

Figure 9.8
A circular staircase from *The Four Books of Architecture* by Andrea Palladio,
1508–1580. (Dover Edition © 1965.)

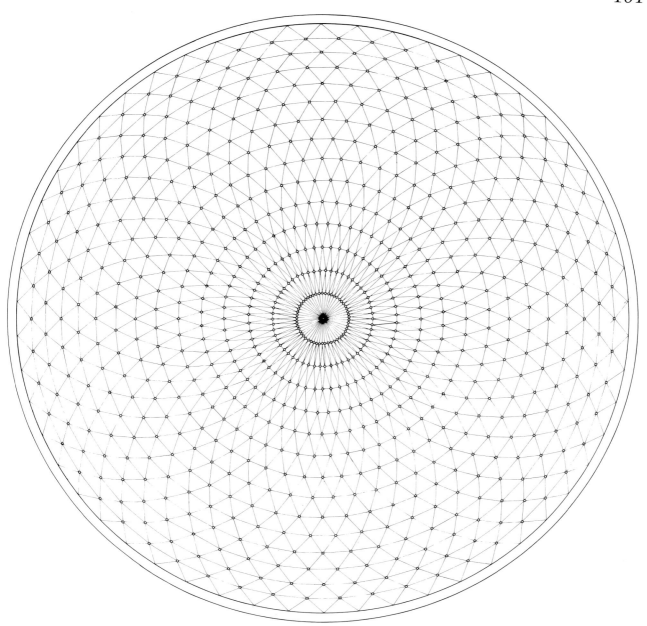

Figure 9.9
Plan of a geodesic dome structure. (From *Geodesics* by Edward Popko, University of Detroit Press, 1968. Reprinted by permission.)

dome structures (Figure 9.9), churches, museums, apartment houses, and parking garages. Numerous patterns (Figure 9.10) can be generated with nothing more than a compass, but the pieces, like a Vasarely painting, rarely will be identical.

Although it is an efficient shape with respect to perimeter, a circle seldom is used as the plan for tall buildings. High winds will be deflected around a circular building rather than down on the sidewalk and, if the building is very tall, the circular shape may reduce the amount of steel required to resist the forces of the wind. However, circular towers have leftover space on the floors, which is difficult to use effectively, are usually

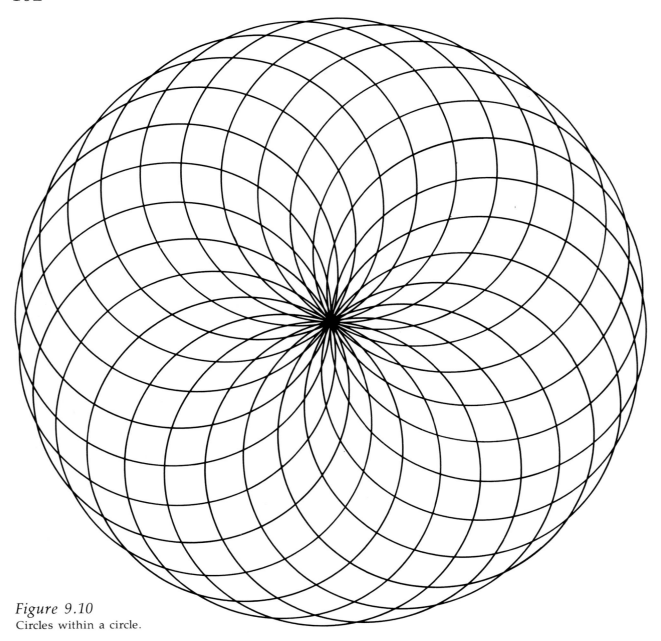

Figure 9.10
Circles within a circle.

costly to construct in comparison to square or rectangle towers, and rarely fully utilize a site. The shape tends toward wastefulness and lacks the great versatility of the rectangle.

THE SEMICIRCLE

Semicircles occur notably in the design of amphitheaters and arches but they also are found in building fronts on curving streets. A straight street can be comprehended in its entirety at a glance and offers few surprises, but a curving street such as Regent Street in London (Figure 9.11) presents an unfolding experience with opportunities for discovery and surprise. The buildings offer a vanishing facade. Complete circles or elliptical shapes such as the Colosseum in Rome (Figure 9.12) with its tiers of small arches literally have no beginning and no ending.

Figure 9.11
The Arc of the Great Quadrant, Regent Street, London by John Nash, architect, planner, and developer. (Original drawing by Thomas H. Shepherd from *London in the Nineteenth Century* by James Elmes, published 1829 by Jones & Co.)

Arches, especially masonry arches, provide a means of creating openings and a means of reducing the amount of stone in walls. The aqueduct of the Pont du Gard, built in the first century A.D. (Figure 9.13) is a prime example. Arches, the rainbows of architecture (Figure 9.14), are unsurpassed as entryways.

The Greek amphitheater (Figure 9.15) accomplished the prime objective of seating a very large number of people as close as possible to the stage. In 350 B.C. there was, after all, no electrical amplification of the human voice. The Greek theater was the theater "au natural." The compact plan wasted not a square inch. The section, too, was remarkable. To this day a steep bank of seats set well back from the stage yields the clearest sight and sound lines possible.

An entire chapter could be written on the geometry of theater seating and, of course, whole books are needed for theater design, one of the

Figure 9.12
Outer wall of the Colosseum, Rome, A.D. 72–80. (Photo by G. E. Kidder Smith.)

Figure 9.13
Pont du Gard, Nimes, France, early first century A.D. (Photo by Jean Roubier, Paris.)

Figure 9.14
Drawing of The Mills Building arch at 220 Montgomery Street, San Francisco, designed by Burnham and Root. The identity of the artist is unknown.

Figure 9.15
The Theater, Epidaurus. c. 350 B.C. To this day, the semicircular Greek theater remains the most effective way to seat a large number of people close to a stage. (Photo by Alison Frantz.)

most challenging of all architectural commissions. Even the controversy of conventional versus continental seating is more often than not decided on the basis of luxurious seating and appearances rather than on the ability of the largest number to see and hear clearly.

Theatrically, the complete circle is ideal for activities such as boxing where the action is revolving and equally, if somewhat randomly, distributed in all directions. Music, drama, and speech require a backdrop and a facing to an audience, hence the semicircle.

ELLIPSES AND SEMIELLIPSES

The history of architecture provides three major works which are elliptical, or nearly so, and all are in Rome. These are the marvelous design in the pavement of the Campidoglio by Michelangelo (Figure 9.16), the Colosseum (Figure 9.17), and the great colonnade of St. Peter's by Bernini (Figure 9.18). A semiellipse can be found in the curving facade of the thirty townhouses at Bath (Figure 9.19).

Ellipses and rectangles are similar shapes, rectangles being flattened squares and ellipses being flattened circles. Both have proportion, a length-to-width ratio which can be varied, and both have a major and a minor axis.

There is an even closer relationship between a rectangle and an ellipse. Every rectangle has associated with it two complementary ellipses (Figure 9.20). The diagonal of the rectangle is the major axis of both ellipses. The length of the rectangle is the minor axis of one ellipse and the width of the rectangle is the minor axis of the other.

Figure 9.16
The elliptical star-patterned pavement in the Piazza de Campidoglio, Rome, designed by Michelangelo in 1536. (Reprinted by permission of the publishers from *Space, Time, and Architecture*, 5th edition by Siegfried Giedion, Cambridge, Mass.: Harvard University Press, 1967.)

The focal points of the two ellipses are the midpoints of the sides of the rectangles. Thereafter, there are some astonishing correspondences.

1. The area of the large ellipse will be to the area of the small ellipse as the length is to the width of the rectangle.

2. The latus rectum of one ellipse will locate the center of the minimum radius of curvature of the other ellipse.

Figure 9.17
The elliptical plan of The Colosseum in Rome. (Reprinted by permission of the publishers from *Space, Time, and Architecture,* 5th edition by Siegfried Giedion, Cambridge, Mass.: Harvard University Press, 1967.)

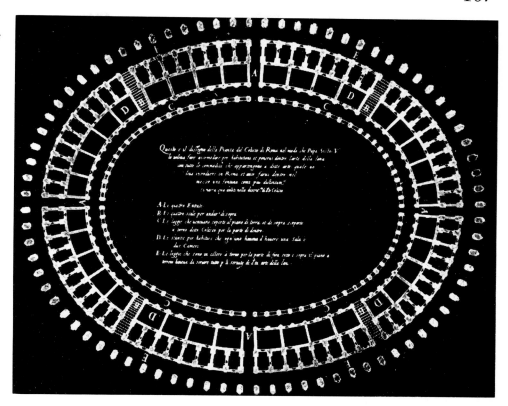

Figure 9.18
St. Peter's Square, Rome. This magnificent elliptical space was designed by Bernini. The colonnade was begun in 1656. (Photo by Donald Ray Carter.)

3. The sum of distances from any point on the ellipse to the foci will be the same for both ellipses and equal to the rectangle's diagonal.

The two ellipses of the square are 45° ellipses (Figure 9.21). The 45° ellipsis stands at the midpoint of all ellipses and is *the ellipse.*

Figure 9.19
Royal Crescent, Bath, by John Wood the younger, 1767–1775. The royal crescent
is a semiellipse and consists of thirty townhouses behind a continuous facade of
giant Ionic columns. (Photo by A. F. Kersting.)

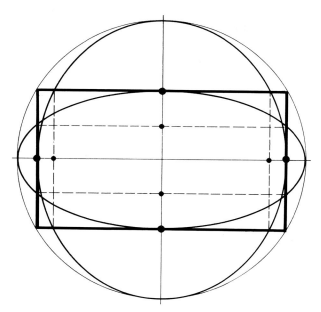

Figure 9.20
The rectangle of the 30° and the 60° ellipse.

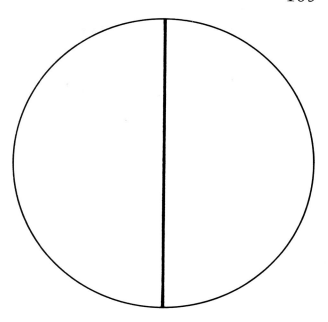

Figure 9.22
The 0° and the 90° ellipse.

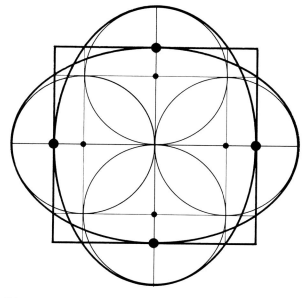

Figure 9.21
A square with its pair of 45° ellipses. The number of geometric relationships resulting from the coupling of a rectangle with its associated pair of ellipses is amazing.

A circle is at the beginning of the series of ellipses and a straight line is at the end of the series. These two shapes (Figure 9.22) are complements and represent as well the extremes in area enclosure, the one having the least and the other having the most perimeter for a given area.

In terms of perimeter, a circle is a more efficient enclosure than a square and, up to a point, an ellipse is more efficient than a comparable rectangle, that is, one with the same length-to-width ratio. When, however, the proportion exceeds 5.75:1, the rectangle is a more efficient enclosure than the comparable ellipse. The 5.75 rectangle and the 5.75 ellipse are shown in Figure 9.23. Both have the same area, perimeter, and length-to-width ratio.

The single focal point at the center of a circle becomes two in an ellipse, offering a unique acoustical experience, since sound originating at one focal point will be concentrated at the other. This very feature, however, renders elliptical rooms nearly useless for normal activities because it would be difficult to hear well at any point other than the focal points. An ellipse is most rewarding as an exterior space and only rarely successful as a room or a building shape. Elliptical mirrors and elliptical pools are, of course, very elegant.

Parabolas and hyperbolas are similar to ellipses. Each of these curves has distinctive focusing and diffusing characteristics which may be useful for particular purposes.

THE INDIANAPOLIS MOTOR SPEEDWAY

Architects are always interested in the shape of things, be it the shape of a vase, a building, or a vehicle. In this instance, the shape of interest is the rectangle of the Indianapolis Motor Speedway (Figure 9.24), a 2.5-mile-long track built in 1912 for automobiles racing at speeds of 100 miles per hour and now used for finely tuned, superpowerful, highly polished machines running at 200 miles per hour—once around the track in 45 seconds! The four turns at the corners are unsuited for the high speeds, yet, in deference to nostalgia, the same drivers, owners, and mechanics who strive for perfection in their machines tolerate an obsolete and dangerous roadway for the performance demonstrations of those machines.

The track, in the first instance, should probably be double the distance around, and probably in the shape of an ellipse, not a rectangle, with properly banked maximum and minimum curves set to attain the best performance of the machines. At the least, such a shape (Figure 9.25) would ensure a graceful race and probably one that is both safer and faster than at present.

Before departing from the fascination of the ellipse, two more illustrations (Figures 9.26 and

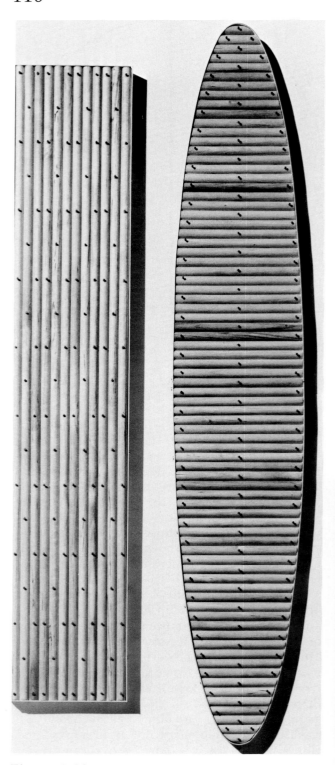

Figure 9.23
The 5.75 rectangle and the 5.75 ellipse. Both shapes have the same area, perimeter, and length-to-width ratio. (Photo by Gerald Ratto.)

Figure 9.24
The rectangle of the Indianapolis Motor Speedway. (Photo courtesy Indianapolis Motor Speedway.)

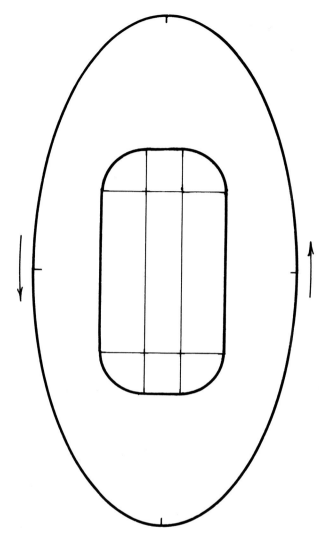

Figure 9.25
A drawing of the Indianapolis Motor Speedway and an elliptical track with twice the circumference drawn at the same scale.

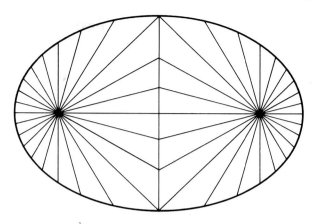

Figure 9.26
The rays of an ellipse. With its two focal points, like two eyes, the ellipse is a captivating shape for graphic design.

9.27) suggest the unique design opportunities present in a shape with not one but two focal points.

SPIRALS

Spirals (along with ellipses) are among the most fascinating forms that can be constructed (Figure 9.28). In nature, the chambered nautilus (Figure 9.29) and the spiral nebula of the universe (Figure 9.30) are the best known examples. In mathematics and geometry, there are numerous spirals (Figure 9.31). In architecture, with the exception of spiral stairs (which actually are helixes), spirals rarely are used, although they convey a powerful sense of growth and progress in time.

Figures 9.32 and 9.33 show an unusual two-story dwelling which conceptually is a spiral. Space literally flows from one room to another and thence to the out-of-doors.

The model in Figure 9.34 is a sketch for an outdoor historical presentation covering 200 years in time. There are twenty wall panels in all, each for a decade, but they are on a logarithmic spiral so the panels become progressively larger with time. The past recedes and the present emerges. The round table is for the future. The design is experimental but serves to illustrate the potential of a spiral as an exhibition hall.

If entry to a spiral is from the outside to the inside, the eye of the spiral becomes extraordinarily important: a glass vault for a crown jewel or a place of internment similar in concept to the entombment function associated with the pyramids of Egypt.

IRREGULAR CURVES

Nature is full of curves and most of them are irregular. They do not follow a line fixed by a formula from geometry but rather take their own

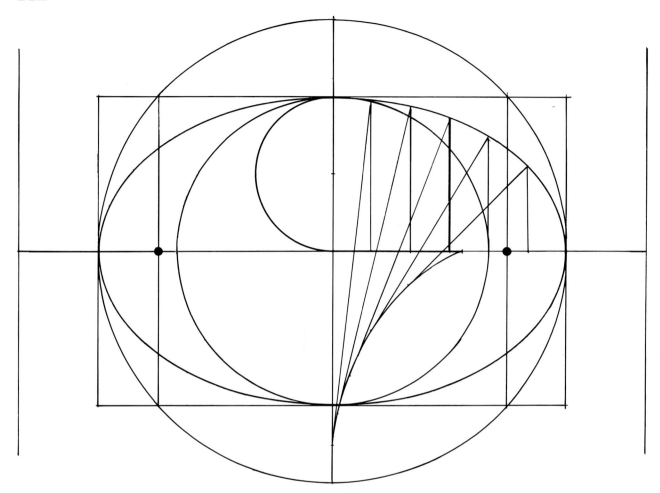

Figure 9.27
Another example of line design with the ellipse.

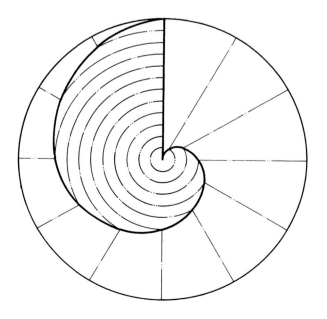

Figure 9.28
The construction of the spiral of Archimedes. A circle has been divided into twelve equally spaced rings and then cut into twelve pie-shaped pieces. The successive intersection of these lines produces the spiral.

Figure 9.29
A chambered nautilus, the classic
example of a spiral in nature. Photo by
Andreas Feininger. (Life Magazine ©
Time, Inc.)

Figure 9.30
Spiral nebula. (Photo by Mount Wilson
and Las Campanas Observatories,
Carnegie Institution of Washington.)

Spiral, hyperbolic or reciprocal

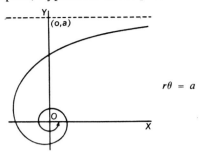

$$r\theta = a$$

Spiral, parabolic

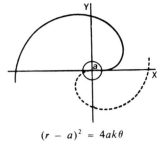

$$(r - a)^2 = 4ak\theta$$

Spiral, logarithmic or equiangular

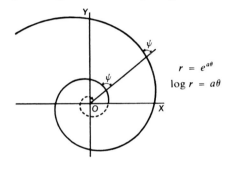

$$r = e^{a\theta}$$
$$\log r = a\theta$$

Spiral of Archimedes

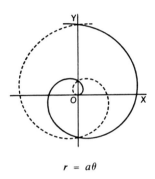

$$r = a\theta$$

Figure 9.31
Four spirals of mathematics. (Reprinted with permission from *Standard Mathematical Tables* by Dr. Samuel M. Selby. Copyright The Chemical Rubber Co., CRC Press, Inc.).

Figure 9.32
Ross House by James Ream, FAIA. In this design the spiral shape allowed the house to unfold from a central circular fireplace. (Photo by the architect.)

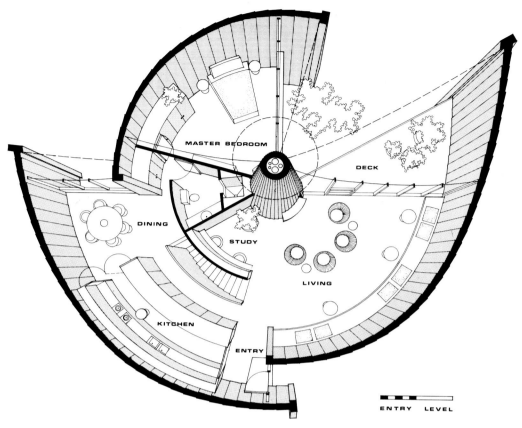

Figure 9.33
Plan view of the Ross House by James Ream showing the entry level. A level below
contains additional bedrooms. The drawing is by the architect.

Figure 9.34
An outdoor spiral
exhibition hall. Each
panel of the spiral
represents ten years
chronologic time on a
logarithmic scale. (Photo
by Gerald Ratto.)

Figure 9.35
Lansdowne Crescent, Bath, 1794. Its serpentine windings follow the contours of
the site. (Photo by Aerofilms, Ltd., London.)

course. Because of the nature of materials and the building processes, it is doubtful that architecture as a regular practice should endeavor to emulate nature on this point, but some designs with irregular curves have proved both buildable and successful. There may, indeed, be circumstances under which they are the only appropriate solution. The serpentine windings of the Lansdowne Crescent at Bath (Figure 9.35) followed, insofar as possible, the natural contours of the site. The

Figure 9.36
Baker House Dormitory, M.I.T., Cambridge, Massachusetts, by Alvar Aalto.
(Photo courtesy M.I.T. Museum and Historical Collections.)

Baker House dormitory at M.I.T. (Figure 9.36) and a high-rise apartment building in Chicago (Figure 9.37) both used reverse curves to good advantage.

Thin sheets of flexible material can be joined together to make lightweight self-supporting wall panels in two distinctive series of irregular curved shapes, one corresponding to the convex aspects (Figure 9.38) and the other to the concave aspects (Figure 9.39) of the regular polygons. The curves, however, are not true circles but rather the result of the bending characteristics of the material held under stress. The strength and stability of the shape is in the curve.

Some of these panels can be fastened together to make curved spaces which may be useful for exhibition and other purposes (Figure 9.40). The fascination lies in creating a nonrectilinear environment where space is flowing and walls are seen as focusing or diffusing elements, not flat planes.

Figure 9.37
Lake Point Tower, Chicago, Illinois, by Schipporeit and Heinrich, 1968. (Photo by Hedrich-Blessing.)

Figure 9.38
Self-supporting curved
shapes. The three-,
four-, and five-sided
shapes are the convex
aspects of the regular
polygons. (Photo by
Gerald Ratto.)

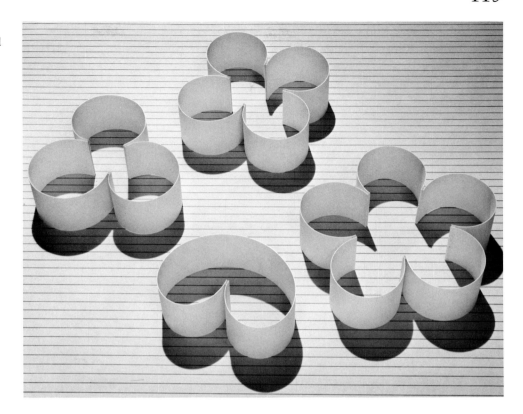

Figure 9.39
Self-supporting curved
shapes. The three-,
four-, and five-sided
shapes are the concave
aspects of the regular
polygons. (Photo by
Gerald Ratto.)

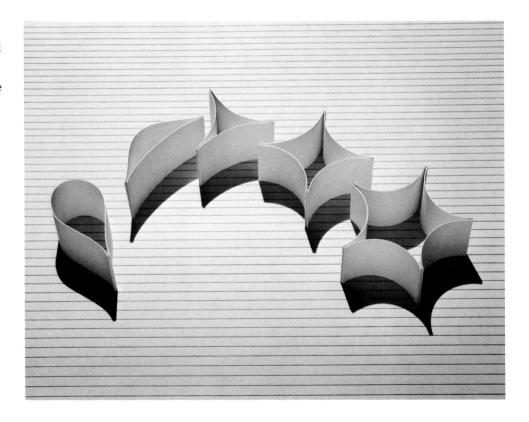

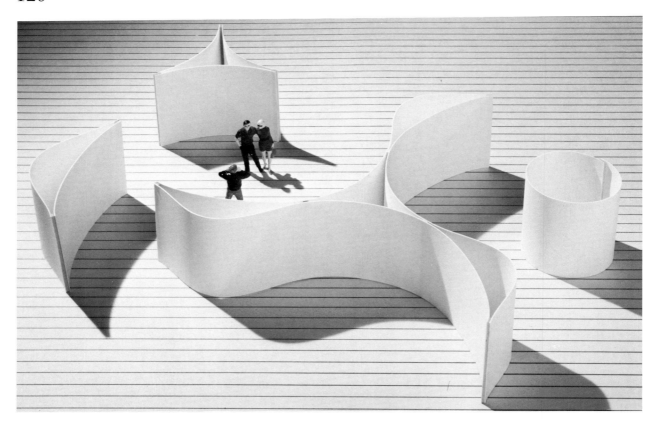

Figure 9.40
Irregular curved panels.
(Photo by Gerald Ratto.)

CHAPTER 10

SURFACE AREA AND VOLUME ENCLOSURE

Figure 10.1
A coin bank by an anonymous designer. At least three
different curvatures contribute to the aesthetic quality
of the piece. (Photo by Colin McRae.)

THE FINELY TEXTURED SURFACE of the flattened spherical coin bank (Figure 10.1) completely encloses a small volume of space, and it does so very efficiently. Judging by the surface area, the amount of material used to make the bank is almost the minimum required for the volume. If the bank had been a cube, for instance, the amount of material would have been greater and the weight would have been greater.

The surface area (and hence the amount of material) required to enclose a given volume is a characteristic of solid shapes just as length of perimeter is a characteristic of plane shapes.

In application, volume/surface and area/perimeter concepts differ in this respect: volume/surface includes the relationship of height to area, whereas area/perimeter assumes height to be a constant.

Very often an architect knows that a room or space will have a flat ceiling of a given height and knows the required floor area. The only variable is the shape as defined by the enclosing walls. Usually the shape will be a rectangle proportioned to meet the functional needs of the room, possibly one of those which Palladio liked. But whether the shape is a rectangle or another plane figure, the geometry of the room is essentially two-dimensional.

When, however, an architect is free to determine the overall shape of a room or building unrestrained by a fixed height requirement, the three-dimensional aspects of geometric shapes become very important. The shapes are solids—prisms, pyramids, antiprisms, and so on—and the essential relationships are volume/surface, not area/perimeter.

THE SURFACE EFFICIENCY OF GEOMETRIC SOLIDS

Of the solid shapes, a sphere encloses a given volume with the least surface area. In this respect, it is the most efficient of the solids. The surface area of all other shapes can be expressed relative to the efficiency of the sphere, taken as 100%. Figure 10.2 shows the surface efficiency of some of the more important series of geometric solids.

The shapes in Figure 10.2 are plotted from left to right according to the number of sides in the base

or equator and from top to bottom according to their surface efficiency. Those with the least surface are closest to the top. The volume enclosed is the same for all of the shapes.

Rectangular prisms, the workhorses of architecture, are not shown, but they fall below the cube and become progressively less efficient as they depart in proportion from the cube. Successive approximations to the sphere, which were described in Chapter 4, begin with the cube and become progressively more efficient as the number of sides in the equator increases.

Antiprisms, prisms, and pyramids are shown at their most efficient proportion, that is, with the one height which reduces total surface area (including the area of the base) to a minimum. The relationship of wall to floor area which produces this unique height is described in the chapters on prisms and pyramids.

The curved line beginning with one sphere at the upper left hand corner of the graph illustrates the dramatic increase in surface area that occurs when a solid shape (in this case a sphere) is divided into two, three, four, or more separate identical shapes. The total volume is the same but the surface area can be increased tenfold or more. The most direct way to increase surface area is to subdivide, and the most direct way to decrease surface is to consolidate.

The fact that a shape is efficient in terms of surface area does not necessarily mean that it is the best functional solution—or the most economical—for a particular purpose. Other considerations may prevail. The goal is to find, on balance, the shape that best satisfies all of the requirements. In practice, the shape of most buildings is determined by program, budget, site, and zoning limitations. Within this framework, the relationship of surface area to volume plays a part.

SURFACE EFFICIENCY OMITTING THE BASE

While a sphere is the most efficient of all enclosures in terms of surface area, a hemisphere is not. When *total* surface area is counted, antiprisms with three or more sides, prisms with six or more

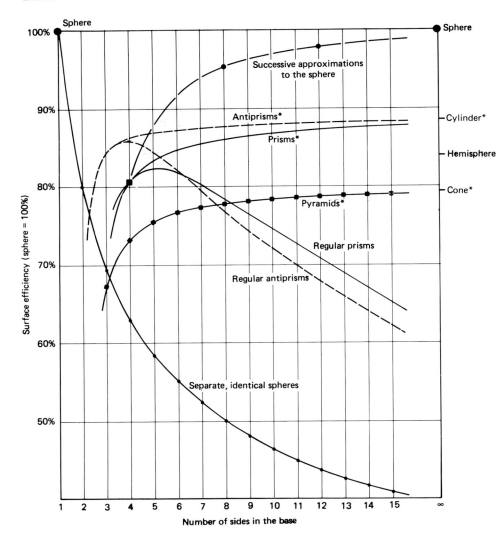

Figure 10.2
The surface efficiency of
various solid shapes.
The series marked with
an asterisk are
proportioned for
maximum volume
enclosure efficiency.

sides, and properly proportioned right circular cylinders enclose a given volume with less material than a hemisphere. Consider that a hemisphere has one-third of its total surface in the area of its base.

More often than not, however, a hemisphere rests on the ground with only the dome exposed to the environment. This leads to a second concept of volume enclosure, one with the area of the base omitted. In this case, a hemisphere is, indeed, the best enclosure when compared to other shapes without their base. This is the concept of Figure 10.3, a graph showing the amount of surface area exposed to the environment of various other shapes relative to the hemisphere. As a complete

enclosure without a base, the sphere is still taken as 100%.

The concept of total surface (Figure 10.2) is important when an object is floating in space and all surfaces are equally weighted; the concept of exposed surface (Figure 10.3) is important when an enclosure is resting on the ground or some other neutral plane. The latter is the situation most frequently encountered in architecture.

Between Figure 10.2 and Figure 10.3 there is a notable change in the order of the curves. For a given volume, antiprisms, prisms, and the right circular cylinder have *less total surface* than pyramids or the cone (Figure 10.2) but when only exposed surface is considered, the order is re-

Figure 10.3
The surface efficiency of solid shapes when the area of the base or ground plane is excluded from the calculations. In this case, the hemisphere is the most efficient of the solid shapes. The series marked with an asterisk are proportioned for maximum volume enclosure efficiency.

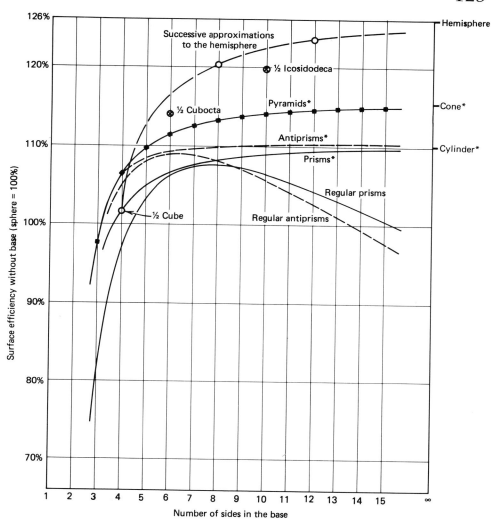

versed and prisms and antiprisms have *more* surface than pyramids (Figure p. 3).

When the area of the base is omitted, pyramids emerge as efficient volume enclosures, presenting a minimum amount of exposed surface to the elements. The idea of huge pyramidal structures under adverse environmental conditions is certainly not an absurdity.

THE EFFECT OF SCALE

There is an important point to remember about both perimeter length and surface area: the relationships between area and perimeter and be-

tween volume and surface change with size. A large cube will enclose more volume per unit of surface area than a small sphere, although the sphere is the more efficient shape.

As Buckminister Fuller liked to point out, a needle will float on water due to surface tension but an exact scale model 100 times larger will sink. The length of longitudinal perimeter subject to surface tension has changed with respect to the volume of the needle. There is now too much weight per unit of perimeter for surface tension to hold and the enlarged needle goes to the bottom.

This is known as the scale effect and leads to the general theorem that the larger an object, the more efficient it will be as a space enclosure. One "supersized" box of breakfast cereal will use a lot less cardboard than four standard boxes containing, in total, the same amount of cereal (assuming the boxes are filled to the top, which is seldom the case). In the same way, the greater the volume of a building, the greater will be the efficiency of the enclosing structure and materials.

A big building will not cost less than a small building. But if four small buildings can be consolidated into one large building, the materials required to enclose the same total volume will be less and the cost per unit enclosed will be less.

This principle has important environmental and energy conservation consequences which tend to weigh against the concept of many small and detached buildings and in favor of fewer and larger structures.

Because of the scale effect, comparisons of perimeter and surface area as between one shape and another are valid only if the shapes are of the same size: the same area in the case of perimeter comparisons and the same volume in the case of surface comparisons.

CHAPTER 11

PRISMS
AND ANTIPRISMS

Figure 11.1
The Baptistry in Florence begun c. 1060. (Photo by Alinari, Editorial Photocolor
Archives.)

PRISMS, as defined in this book, are regular polygons and other two-dimensional figures with the added dimension of thickness or height. The walls are perpendicular to the base and the volume is simply the area of the base times the height. Floor tiles are prisms, boxes are prisms, most rooms and nearly all buildings are prisms. The two-tiered Romanesque Baptistry in Florence (Figure 11.1) in geometric terms is an octagonal prism with a pyramidal roof.

Since the base can be any imaginable closed shape and the height can range from near zero to infinity, there can be an immense number of different prisms. The walls usually are rectangles, but the circle of the regular polygon series can be the base of a prism, and the concave and convex aspects of the regular polygons and other curved shapes can also be prisms. The cylindrical chapel at M.I.T. (Figure 11.2) and the half-cylindrical glass roof of the Crystal Palace (Figure 11.3) are prisms.

There is one important division within the group of prisms. *Regular* prisms have square sides and height is not a variable. In all other prisms, including semiregular, rectangular, and cylindrical, height is a variable. The twin towers of the World Trade Center (Figure 11.4) are majestic semiregular prisms.

THE REGULAR PRISMS

Figure 11.5 shows the regular prisms of two, three, four, five, six, and eight sides. The best known is the cube. Regular prisms with a number of sides evenly divisible by four are the midsection of the solid series of shapes successively approximating the sphere (Chapter 4).

Like the solids, the whole shape is determined by one dimension. Height is not a variable but is always the length of one side. As the number of sides increases, the height decreases with respect to the area of the base. A regular prism with a great many sides resembles a flat circular disc.

Of the regular prisms, the pentagonal prism (not the cube) offers the best volume enclosure when total surface area is important. The octagonal prism is the most efficient enclosure when only exposed surface, the walls and the roof but

not the floor, is important. Curiously, a regular hexagonal prism and an octahedron enclose a given volume with exactly the same amount of surface area.

MINIMUM SURFACE PRISMS

In all but the regular prisms, height is a variable. The unique prisms are those in which the height is proportioned to the base so as to minimize the surface area.

Of the rectangular prisms, a cube is the most efficient volume enclosure. The more a rectangular prism departs from the proportions of a cube, the greater will be the surface area needed to enclose a given volume.

The area of the walls of a cube is four times the area of the base. This is plain to be seen. It is astonishing, however, to find that this simple relationship between wall area and floor area applies not only to the cube, the most efficient of the rectangular prisms, but to *every* prismatic solid at its most efficient proportion.

The shape of the base may be a rectangle of any proportion, a regular or irregular polygon, a concave or convex aspect of the regular polygons, a circle, a semicircle, ellipse, irregular curve, or any other enclosed shape, but *when the area of the walls is four times the area of the base, the total surface required to enclose a volume will be the least possible.* The shape will be at its most efficient proportion in terms of volume enclosure.

When only exposed surface is important, the area of the roof and the walls but not the floor, a similar relationship exists. *The exposed surface of any prism will be minimal when the area of the walls is twice the area of the floor.* Thus a half-cube is the best of the square rectangular prisms when only exposed surface is counted.

Since the wall area of a prism is the perimeter of the base times the height, the height required for optimum volume enclosure can easily be found for any prismatic shape.

In the case of semiregular prisms which have a base in the shape of a regular polygon, the area of the walls will be four times the area of the base when the height is equal to the inside diameter of the base. Total surface will be minimal. The area of

Figure 11.2
The cylindrical brick chapel at M.I.T. by Eero Saarinen, 1955. (Photo by Joseph
W. Molitor.)

Figure 11.3
An interior view of The Crystal Palace by Sir Joseph Paxton, 1851. The roof of the gallery is a half-cylinder in glass. (Lithograph by Joseph Nash, Victoria & Albert Museum, London, Crown copyright reserved.)

Figure 11.4
The World Trade Center in New York by Minoru Yamasaki in association with Emery Roth and Sons. (Photo by Joseph W. Molitor.)

Figure 11.5
The regular prisms of
two, three, four, five,
six, and eight sides.
(Photo by Gerald Ratto.)

the walls will be equal to twice the area of the base when the height is equal to the inside radius of the base. In this case, exposed surface will be minimal. Figure 11.6 show the series of semiregular prisms with the height equal to the inside radius of the base. Figure 11.7 shows an enlargement of the pentagonal prism as a room with this proportion.

The optimum height for rectangular prisms of any floor proportion is also easily found. The area of the walls will be twice the area of the base when the height is the length times the width divided by the length plus the width. With this height, the exposed surface will be the least possible. The area of the walls will be four times the area of the base when the height is twice the above. In this case, total surface will be minimal.

Other prismatic shapes have different formulas, some more complicated than those for rectangular and semiregular prisms, but the relationship of wall area to floor area is always the same for minimum surface enclosure.

THE PARTHENON AND THE "GOLDEN" ROOM

The greatest example of the Classical phase of Greek art in full maturity (according to H. W. Janson, the author of *History of Art*) is the Parthenon (Figure 11.8), a white marble structure on the most prominent site of the Acropolis overlooking Athens. It is considered the perfect embodiment of Classical Doric architecture and remains the best example of a rectangular prism to be found in architecture.

The building has been analyzed in every possible way and there are numerous explanations for

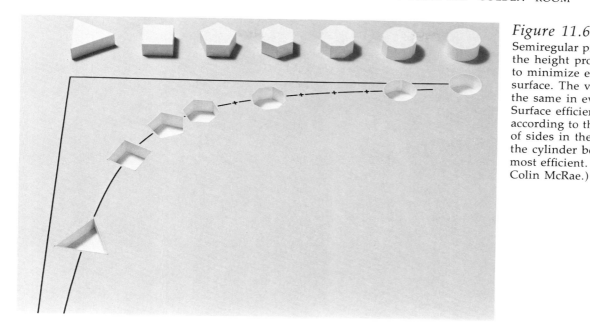

Figure 11.6
Semiregular prisms with the height proportioned to minimize exposed surface. The volume is the same in every case. Surface efficiency varies according to the number of sides in the polygon, the cylinder being the most efficient. (Photo by Colin McRae.)

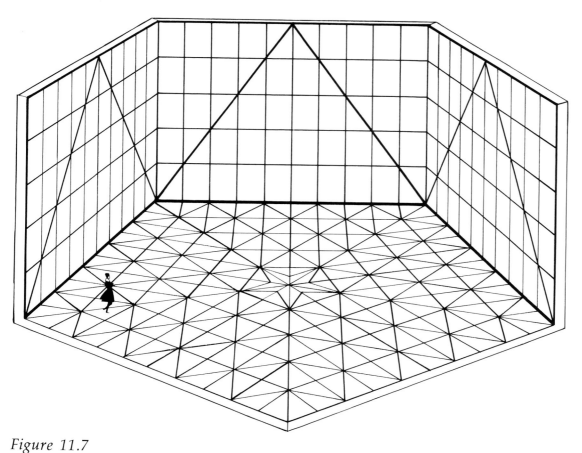

Figure 11.7
A diagram of a pentagonal room with the height of the walls equal to the inside radius of the room. With this height, the modular relationship of floor and walls is the same and the surface area is minimal.

Figure 11.8
The Parthenon in Athens, built in 448–432 B.C. As in many buildings today,
there is an even number of columns on the front so that a space, rather than a
line, will be at the center. (Photo by Alison Frantz.)

its immense appeal. These explanations are often
given in terms of mathematical ratios and proportions, an emphasis which may be exaggerated.
The architect of the Parthenon, like many architects of today, probably followed his own inner
rules and no one can now know exactly what those
rules were. Needless to say, he made the most of
what he had to work with!

Louis Sullivan in his *Kindergarten Chats* noted
that logic has its excellent uses. At least some of
the principles formulated in this book are known
intuitively. The mathematics serve to reinforce
and render more exact this intuitive knowledge
but will not carry the architect the whole way in
the design of a completely satisfying edifice. At
times, logic may suggest for consideration a particular solution as a starting point and, for that
reason, it is useful.

An example is the "golden" room, a rectangular
prism with a floor in the ratio of the golden mean,
ϕ. When the area of the four walls is made equal to
twice the floor area, the relationship which yields
the greatest volume per unit of wall and ceiling
area for prismatic solids, the height of the room
will be 0.618 (the reciprocal of ϕ) times the width
of the room. If the room width is one, the volume
will be one and the length of the great diagonal
will be two.

This room is shown in Figure 11.9. For anyone
intrigued by the aesthetics of numbers in
geometry, the golden room, like a rich dessert, is a
source of immediate gratification. As judged from
the drawing, the proportions of this room are
equally rewarding.

The end wall of the room is itself a golden rectangle. To this has been added the spiral of the
golden mean, a construction of quarter-circles
within a progression of squares.

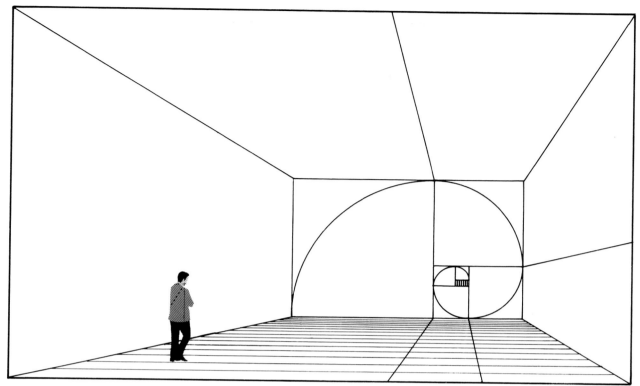

Figure 11.9
The "golden" room.

ANTIPRISMS

Antiprisms (Figure 11.10) are unique shapes seldom used by architects although they are surprisingly efficient volume enclosures with a rigid triangular wall structure.

The exterior walls of antiprisms are isosceles triangles while the walls of prisms are rectangles. In spite of this difference, prisms and antiprisms have many similar characteristics.

Both fit between parallel planes, have tops and bottoms of identical polygons, and are identified by the number of sides in the top or bottom. Regular antiprisms, those with equilateral triangular sides, and regular prisms, those with square sides, have much in common with the classical solids. The faces are regular polygons, the corners are all alike, they can be inscribed in a sphere, and they follow Euler's formula linking the number of sides, vetrices, and faces.

Because the sides of the antiprisms are isosceles triangles rather than rectangles, the top and bottom polygons, while parallel, are rotated a half-phase with respect to each other. Unlike prisms, antiprisms have a built-in twist or turn which is especially noticeable in the upward spiral lines of an antiprism column (Figure 11.11). Like diagonal bracing, these spiral lines give the shape extra resistance to the twisting forces of wind and earthquake. Antiprisms and modified antiprisms also can be used to construct near-perfect domes with similar strength characteristics.

Because of the triangular construction of the walls, antiprisms always have an even number of wall panels and an equator with twice as many sides as there are in the base.

The walls of an antiprism slope. They are not upright, which is one reason why, by definition, they are not prisms and why they are seldom used architecturally. Like the surface of the classical solids, the walls of antiprisms face to a point at the center of the shape. The walls of a prism face not to a point but to an axis at the center.

Antiprisms are pyramidal structures. The volume is not simply the area of the base times the height (another reason why they are not prisms)

Figure 11.10
Regular antiprisms of two, three, four, five, and six sides. The sides are
equilateral triangles. (Photo by Gerald Ratto.)

but is found by summation of the internal
pyramids, or the prismoidal formula.

THE FIRST FIVE REGULAR ANTIPRISMS

The first five in the series of regular antiprisms are
noteworthy.

First is the tetrahedron, an antiprism with a
base polygon of two sides, the beginning of the
series of regular polygons. The top and bottom
lines are turned 90° with respect to one another.

Second is the octahedron, although to qualify as
an antiprism it must rest on a face and not stand
on a point. In this position, the equator of an
octahedron is a hexagon.

Third is the square antiprism, the most efficient
of the regular antiprisms in terms of the total
amount of material needed to enclose a given volume. In this respect, the square antiprism is
superior even to a cube.

Fourth is the pentagonal antiprism, the midsection of the icosahedron, a classical solid of twenty
equilateral triangular faces.

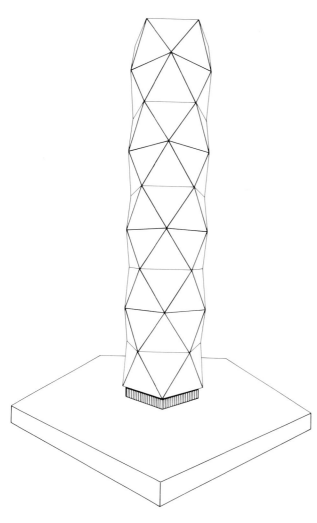

Figure 11.11
A column of pentagonal antiprisms. The spiral construction adds to the strength of the column.

Figure 11.12
Plan view of the five principal antiprisms.

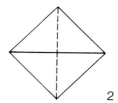

2

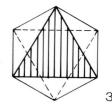

3

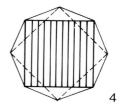

4

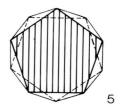

5

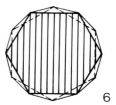

6

Finally, the hexagonal antiprism is noteworthy because it is the most efficient volume enclosure when only exposed surface is counted, that is, when the area of the base or ground plane is excluded from the calculations.

Figure 11.12 shows the plan view of these five antiprisms.

When the number of sides is small (six or fewer for total surface and eight or fewer for exposed surface) regular antiprisms are better volume enclosures than regular prisms with the same number of sides. Beyond that, regular prisms are the best choice between the two for efficient volume enclosure. These series of shapes are shown as curves in Figures 10.2 and 10.3.

SEMIREGULAR ANTIPRISMS

The walls of regular antiprisms are equilateral triangles and height is not a variable, but the walls of semiregular antiprisms can be isosceles triangles of any proportion and height is a variable. As with prisms and pyramids, however, there is only one proportion which, for a given number of sides, will reduce to an absolute minimum the surface area required to enclose a given volume.

However, unlike prisms and pyramids, the height of an antiprism for optimum volume enclosure is not a constant relationship between the area of the walls and the area of the base, but rather a relationship which varies with the number of sides in the antiprism. This varying relationship is unique to antiprisms, making them at once more complex and more interesting than many of the geometric shapes regularly used by architects.

As the graphs in Chapter 10 show, antiprisms with a height calculated to minimize surface area are always more efficient volume enclosures than prisms with the same number of sides. The greatest difference occurs when the number of sides is relatively small, from two to eight or so. As the number of sides increases, prisms and antiprisms with minimum surface come closer together. Ultimately, both become right circular cylinders.

CHAPTER 12

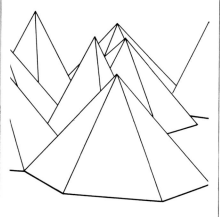

PYRAMIDS AND CONES

Figure 12.1
The Pyramids of Mycerinus, Chefren, and Cheops at Giza, c. 2500 B.C. (Photo by
Hirmer Fotoarchiv, Munich.)

THE MASTER BUILDERS OF EARLY CIVILIZATION, the Egyptians, developed pyramid construction to its ultimate in the fourth century B.C. The triad of grand structures they built in Giza (Figure 12.1) are the most recognized of all pyramids and rank among the seven wonders of the ancient world.

Numerous books have been written about the pyramids of Egypt, their significance, and the people who built them. Interestingly, historians now believe these funerary monuments were not built on the backs of toiling slaves but by workers who were paid for their labor.

Although they may have served a government purpose as vast public works projects, the primary function of the Egyptian pyramids was not to serve the living but to be timeless repositories for the bodily remains and wordly possessions of the dead. As such, their interior space, compared to their bulk is almost nil.

In contrast, pyramids today are proposed by architects and builders to enhance livability and to provide exciting new designs for interior space (Figure 12.2). These pyramid structures are constructed not of stone but of exceptionally lightweight space-frame materials that can be assembled easily into modular units (Figure 12.3).

Unlike Egyptian pyramids, these pyramids have extensive interiors (Figure 12.4). Many levels on the outside walls can be used for individual apartments, while the huge, column-free interior space can be devoted to indoor gardens and recreation (Figure 12.5). These apartment pyramids also may have, at every level, open exterior balconies that do not cast shadows on those below.

Pyramids, like prisms, can have any number of sides, from two to infinity, and can be of any height. Pyramids with a square base are particularly useful for architectural structures. Although the walls slope, the floor plans on every level are square and fit the near-universal system of rectangular interior spaces, materials, and furnishings.

MINIMUM SURFACE PYRAMIDS

The volume of most architectural shapes is found simply by multiplying the area of the base times the height. The volume of pyramids and cones, however, is found by multiplying the area of the base times one third of the height, a relationship proved by the Greek geometer and astronomer Eudoxus of Cnidus (408–353 B.C.).

Based on the relationship of height to base, there are four series of pyramids of special interest to the architect. In all four, the base is a regular polygon of any number of sides.

Figure 12.2
A contemporary concept of pyramidal buildings at Las Vegas. (Illustration courtesy of Carlos Diniz Associates, Los Angeles.)

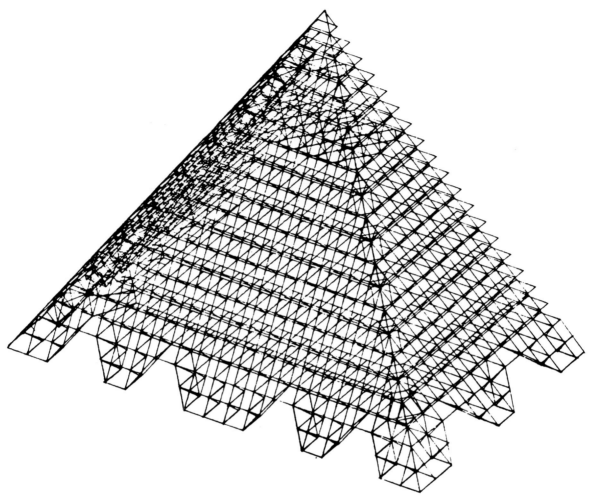

Figure 12.3
A computer drawing of the structural elements of a pyramid. In contrast to the pyramids of Giza, the structure is very light in weight. (Courtesy Leon Nadalski and Istvan Kadar, Bechtel Civil & Minerals, Inc.)

The first series consists of pyramids which can be inscribed in a hemisphere and have a ridge slope of 45°. In these, the height is equal to the *outside* radius of the base. The half-octahedron is an example.

The second consists of pyramids in which the height is equal to the *inside* radius of the base. In these, the walls, not the ridge, slope at an angle of 45°. The pyramid of the cube is an example.

The two remaining series are minimum surface pyramids, one with minimum *total* surface for a given volume and one with minimum *exposed* surface for a given volume.

Eudoxus himself would have been pleased to

discover that the volume enclosed per unit of total surface will be the greatest when the area of the walls of a pyramid is equal to three times the area of the base. To obtain this relationship, the height of the pyramid must be the square root of two times the inside *diameter* of the base. The slope of the walls will be 70° 32', the steepest of the four series, close to that of the pyramid of Caius Cestus (Figure 12.6).

The fourth series of pyramids is the most important of all to architects. This series assumes that the pyramid is on the ground and that the only exposed surface is the area of the outside walls.

Regardless of the number of sides, the exposed

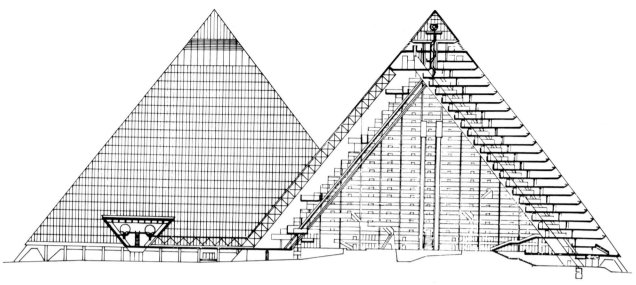

Figure 12.4

A proposal for housing 2000 people in pyramid-shaped buildings twenty floors in height. The interior includes extensive gardens and recreation areas for use the year round. Designed for Siberia by A. and E. Schipkova in 1967. (From *Architecture d'Aujourd'Hui*, No. 134, October 1967.)

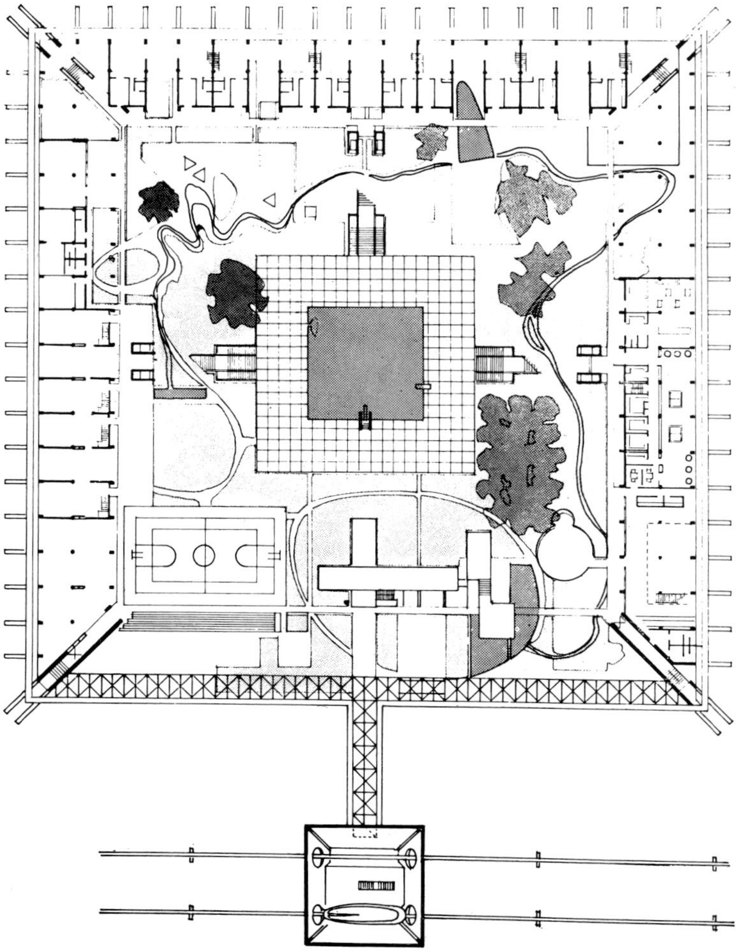

Figure 12.5

A plan view of the interior of the Schipkova proposal.

Figure 12.6
The pryamid of Caius Cestus, Rome, 15 B.C. (Reprinted from *Architecture in Space* by Bruno Zevi, © 1974, by permission of the publisher, Horizon Press, New York.)

Figure 12.7
A home with a pyramidal roof in Colon by Ernst Nolte. (From *Das Eigene Heim* by Guido Harbers, © 1951 by Otto Maier Verlag, Ravensburg, Federal Republic of Germany.)

surface for a given volume in this series will be minimum when the area of the outside walls equals not three times the area of the base but the square root of three times the area of the base. This is a curious, but nonetheless exact, relationship.

When the base is a regular polygon, this relationship will occur when the height is equal to the square root of two times the inside *radius* of the base. The slope of the walls will be 54° 44′. The interior pyramids of the octahedron have this wall slope.

While the ancient Egyptians are credited with having superb mathematical capabilities, the pyramids of Giza do not satisfy this or any other height-to-base relationship that might be expected. The Great Pyramid of Cheops, largest of the Giza pyramids, has a wall slope of 51° 6′.

When total surface is considered, prisms enclose a given volume with less material than do pyramids. When the surface of the base or ground plane is excluded from the calculations, however, pyramids enclose a given volume with the least surface. This is a great virtue of pyramids as space enclosures, especially in a very hot or cold climate where extensive insulation is required. Pyramids deflect wind upward rather than down on the street and they cast a very slight shadow. However, they require more ground area than a prism of the same volume.

In addition to their distinct architectural advantages, theories abound as to the metaphysical attributes of pyramids. It is claimed that pyramids mummify organic matter; that they maintain the sharpness of a razor blade inserted inside, even if made of cardboard or Styrofoam; that when used as hats, they will relieve headaches. Pyramids have been identified as cosmic generators as well as resonators for the energy of the cosmos. All of these theories have fascinating appeal, but proof of their validity lies in a realm beyond that of architecture.

PYRAMIDAL ROOFS

Pyramidal roofs such as the one on an Ernst Nolte house in Cologne, Germany (Figure 12.7), are used to shelter square or rectangular dwellings in spite of the fact that they do not function well in areas of heavy snowfall. The problem is that pyramidal roofs dump snow evenly on all sides, including the front entrance area, where it may prove unwanted, and the north side, where it will be slow to melt.

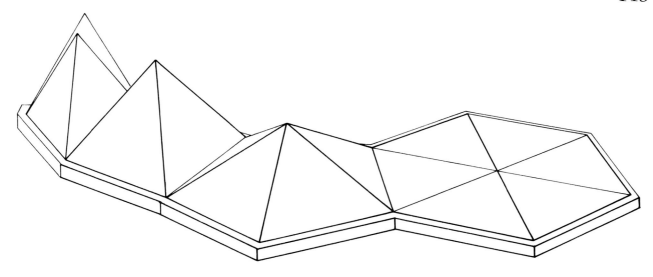

Figure 12.8
The five perfect pyramids.

Roof shapes of a serious slope should be selected which direct snowfall where buildup on the ground will be harmless and where the warm rays of sun will be most effective in reducing the impact of the snow. There are at least a dozen basic geometric roof shapes and countless variations to choose from. As roof shapes, many of these have functional, economic, and aesthetic advantages over the pyramidal shape.

THE FIVE "PERFECT" PYRAMIDS

Equilateral triangles can be joined at a point with sides in common to make the five "perfect" pyramids (Figure 12.8).

The middle pyramids are three-fourths of a tetrahedron, one-half of an octahedron, and one-fourth of an icosahedron, the three Platonic solids of the equilateral triangle.

To this 3–4–5 series, two more can be added: the five-sided pyramid can be flattened into a six-sided pyramid, a hexagon, and if it is agreed that a line is a two-sided figure, another pyramid, one with two equilateral triangular faces, can be added at the beginning.

Thus 20 equilateral triangles can be joined to form a sequence of five pyramids with two, three, four, five, and six sides. The dimensions are shown in Figure 12.9. In this sequence, the beginning and end shapes, the line and the hexagon, have no volume except the thickness of the material.

THE INTERIOR PYRAMIDS OF THE CLASSICAL SOLIDS

The five regular and the thirteen semiregular classical solids are pyramidal structures. This fact is apparent in the calculation of the volume of the more complicated solids. Antiprisms also are comprised of pyramids.

The interior pyramids of the geometric solids are the reverse of the Egyptian monuments. In Egyptian pyramids, the walls are visible but the base is not. In pyramids of the solids, the base is visible as a face of the solid, but the walls of the pyramid are invisible. These interior pyramids of the solids intersect at the geometric center of the solid, the center of the enclosing sphere.

One solid may have two or three different kinds of pyramids (square, triangular, pentagonal, etc.) but the walls of the pyramids will always be identical isosceles triangles. These walls are the interior partitions of the solid and are illustrated in Chapter 13.

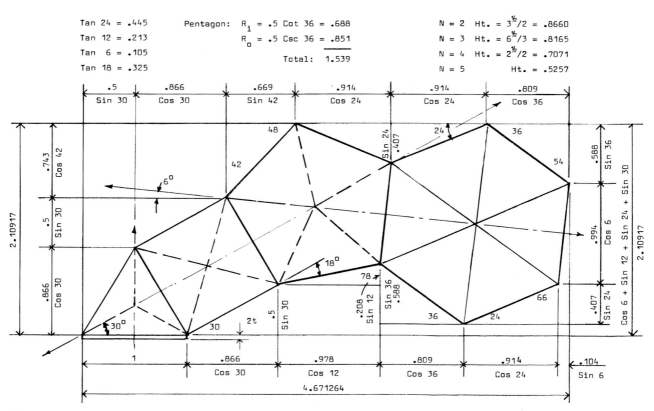

Figure 12.9
A dimensioned plan of the five perfect pyramids.

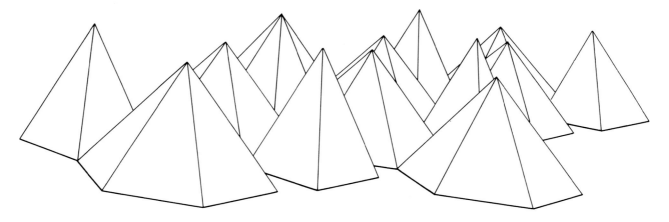

Figure 12.10
The fourteen pyramids of the truncated octahedron. There are two different kinds
of pyramids but the sides are identical.

Figure 12.10 shows six square and eight hexagonal pyramids, all with identical sides. When folded together, this village of fourteen "tepees" will make a truncated octahedron, one of the eighteen classical solids. All of the classical solids can be assembled from similar pyramidal "villages."

An actual American Indian tepee (Figure 12.11) resembles a right circular cone but is in fact a

Figure 12.11
A Sioux Indian tipi. (From *The Indian Tipi* by Reginald and Gladys Laubin, © University of Oklahoma Press.)

pyramid with a polygonal base of eighteen or more sides. The skins are stretched between the poles and the sides are flat. Most interesting about the Sioux tepee (or tipi, as used by the respected authors) is the fact that the shape on the ground is not a perfect circle or regular polygon but an egg shape (Figure 12.12).

RIGHT CIRCULAR CONES

As the number of sides increases, pyramids begin to resemble cones. Properly proportioned, that is, with a height that will assure either minimum total surface or minimum exposed surface, a cone is the most efficient of the pyramidal shapes.

For a given volume, the total surface of a right circular cone, including the base, will be minimal when the height is equal to the diameter of the base times the square root of two.

Excluding the area of the base, the surface will be minimal when the height is equal to the radius of the base times the square root of two.

Compared to cylinders, cones are more efficient when exposed surface is important but they are less efficient when total surface, including the area of the base, is counted.

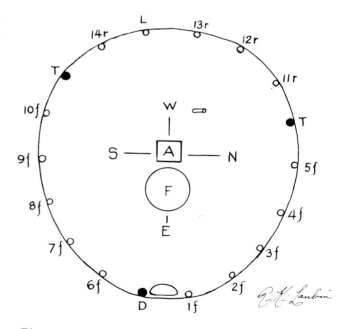

Figure 12.12
Plan view of the Sioux tipi. The arrangement of the eighteen poles results in several different curvatures at the base. It is not a perfect circle. (From *The Indian Tipi* by Reginald and Gladys Laubin, © University of Oklahoma Press.)

CHAPTER 13

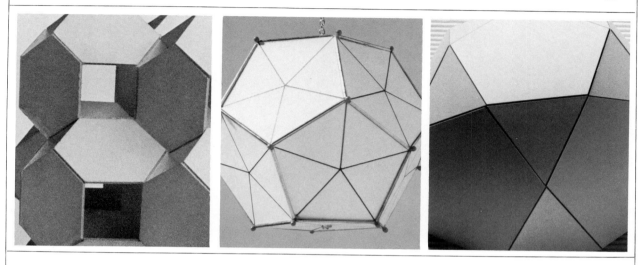

THE CLASSICAL SOLIDS

Figure 13.1
Eclipse, a three-story metal sculpture in the lobby of the Hyatt Regency Hotel, San Francisco, by Charles Perry. (Photo by Balthazar Korab.)

THE GEOMETRY OF THE CLASSICAL SOLIDS is the basis for the design of soccer balls, world maps, portable shelters, massive spherical pendentives, lacy chandeliers, intricate sculpture, huge domes, and numerous other space enclosures and objects.

On the surface, these solids are assemblies of the primary and secondary regular polygons—squares, pentagons, and so on. Beneath the surface, they are assemblies of pyramids, a fact that becomes apparent in the calculation of volume.

There are eighteen in all, five regular (Platonic) and thirteen semiregular (Archimedian) solids. The five regular solids and seven of the thirteen semiregular solids are closely linked by the process of truncation. The remaining six assemblies, including the great rhombicosidodecahedron (rhombicosi), have surface polygons with a length of side in common but not linked by the process of truncation.

The classical solids have distinctive shape, size, and position but are without proportion. Like regular prisms and regular antiprisms, the whole shape is determined by just one dimension, usually the length of an edge. Size is the only variable. The classical solids may be big or small, opaque or transparent, but they cannot be high or wide or long or narrow.

In spite of this limitation, the architectural characteristics of the solids are worth review. Who is to know how they may be used in the future?

GEOMETRIES OF THE SOLIDS

The points or vertices of the solids touch the surface of an imaginary sphere enclosing the shape. This is the important characteristic Buckminster Fuller used to convert the surface of an icosahedron to the marvelous geodesic dome structures of today. By projecting natural subdivisions of the twenty equilateral triangular planes of the icosa to the surface of a sphere, Fuller evolved the intricate, interlocking, modular construction of his famous domes.

The eighteen solids are also approximations to a sphere in terms of surface efficiency. Thirteen of them have the characteristic compactness of a sphere. The snub dodeca, one of the six remaining solids, is the most compact of the solids. Figure 13.2 shows the surface efficiency of the classical solids relative to the sphere at 100%. At the bottom of the scale is the tetrahedron, a four-sided shape with more surface for a given volume than any other solid.

All of the solids can be inscribed in a sphere. Some of them, notably those in the cube-octa series and the great and small rhombicosi, will also fit snugly into a cube. Shapes that fit well in a cube retain a clear relationship with the Cartesian system and thus are found in architecture more frequently than other shapes. Parallel surfaces between an object and the four walls, floor, and ceiling of a room is one visual link which stabilizes a shape in space.

PYRAMIDAL CONSTRUCTION

The five regular solids are assemblies of identical pyramids of three, four, or five sides corresponding to the polygons on the surface. The pyramids meet at the center, and the number of them is the same as the number of faces in the solid—four, six, eight, twelve, or twenty.

Semiregular solids have two or three different polygons on the surface and hence have two or three differently shaped pyramids meeting at the center. Although in any one solid the pyramids of construction have a different number of sides corresponding to the surface polygons, the sides themselves are identical isosceles triangles. These isosceles triangles (Figure 13.3) are the interior partitions of the classical solids.

The interior partitions of the truncated octahedron are seen in Figure 13.4. As a method of partitioning space with respect to a point source, the classical solids offer an alternative to the standard Cartesian system. Only the pyramidal structure of the octahedron divides space into the mutually perpendicular planes of the Cartesian system. The pyramidal construction of the other solids divides space in other ways and in other proportions. For instance, a cubocta divides space into fourteen parts, 40% (by volume) in triangular pyramids and 60% in square pyramids.

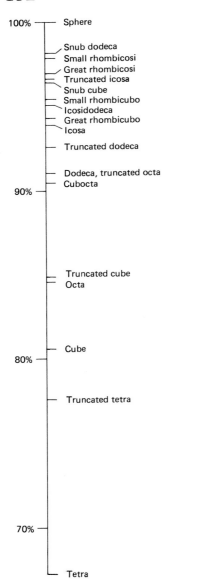

Figure 13.2
The surface efficiency of the classical solids. The shapes with the least surface for a given volume are closest to the top and those with the most surface are nearest to the bottom.

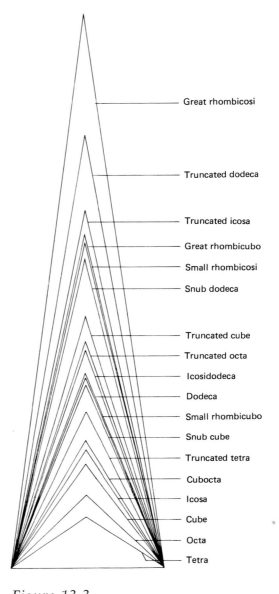

Figure 13.3
The isosceles triangles of the interior partitions of the classical solids.

SOLIDS LINKED BY THE PROCESS OF TRUNCATION

Twelve of the solids are linked by the process of truncation and fall into three distinct groups, the tetra-octa, cube-octa, and icosa-dodeca. These groups correspond to the 3/6, 4/8, and 5/10 shapes in plane geometry.

First of these is the tetra-octa series (Figure 13.5), which begins with the **tetrahedron.** A tetrahedron is one of the five Platonic solids, one of the five "perfect" pyramids, and the first in the infinite series of regular antiprisms. As a pyramid itself, the tetra is found in the construction of the cubocata.

Of the eighteen classical solids, the tetra has the

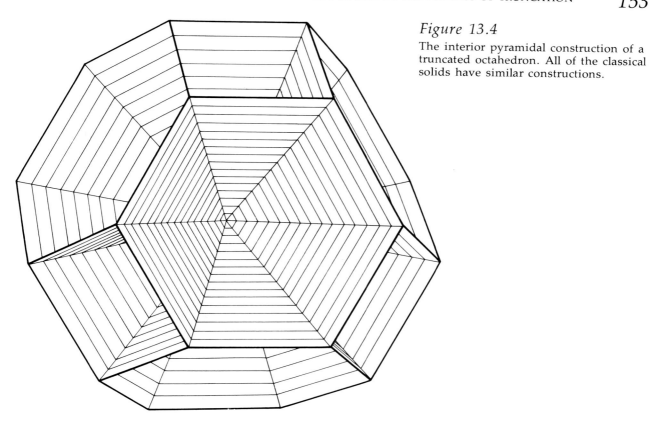

Figure 13.4
The interior pyramidal construction of a truncated octahedron. All of the classical solids have similar constructions.

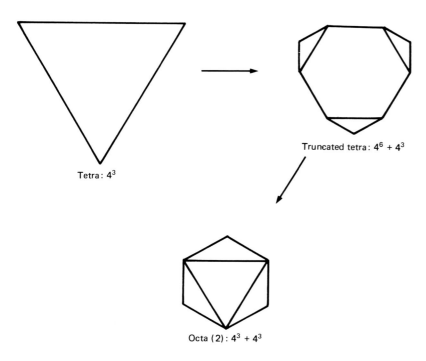

Figure 13.5
Plan view of the tetrahedron, truncated tetrahedron, and small octahedron linked by the process of truncation. Smooth transitions can be made between one shape and another.

Tetra: 4^3

Truncated tetra: $4^6 + 4^3$

Octa (2): $4^3 + 4^3$

largest surface area for a given volume but the fewest number of faces, edges, and vertices. It is the simplest to construct; the four faces can be cut from a single equilateral triangle.

In terms of structural framing, it is the strongest of the geometric solids. Its triangular sides are rigid frames. For this reason, the tetrahedron often provides the geometric basis for space frames and other lightweight tubular constructions that can be extended far into space.

Next after the tetra in the tetra-octa series is the **truncated tetrahedron,** which combines large hexagons with an equal number of small equilaterals. The truncated tetra is the first of the thirteen semiregular (Archimedian) solids. The tetra and the truncated tetra are the only solids whose principal faces meet at acute angles. By surface area, the truncated tetra is six parts hexagon and one part triangle; by volume, it has eighteen parts in hexagonal pyramids and five parts in triangular pyramids.

If truncation of the tetrahedron continues, the end result is a **small octahedron,** another of the Platonic solids.

CUBE-OCTA SERIES

The second group of solids linked by the process of truncation is the cube-octa series (Figure 13.6). It begins with the cube.

A **cube** or **hexahedron** is familiar from childhood as the shape of alphabet blocks which could be put together in countless ways, knocked down, and put together again in a new way. It is tempting to say that a cube is also the building block of architecture with the same limitless variety and flexibility as toy blocks. It is not.

Buildings are rarely in the shape of cubes. When they are, it is often by coincidence and not by design. For instance, a height limit may happen to be the same as the dimensions of a site. The proportions of a cube have few functional advantages for most building types and the aesthetics of a cube are satisfying only under special conditions. It is rare also that the height of a room will equal its width and length, although small rooms—elevators, for instance—come close.

The shapes and spaces of buildings are not made by assembling cubes but by the construction of structural frameworks and the application of materials to them. As a system to assure modular coordination, the square grid is unsurpassed, but a three-dimensional grid of cubes (Figure 13.7) is rarely used.

Of the rectangular prisms, the cube is the most efficient volume enclosure when all of the surfaces are considered—floor, walls, and roof—but when compared to other classical solids, the cube is relatively inefficient, exceeding only a tetra and a truncated tetra.

The six square faces of a cube can be truncated to produce a new shape which retains all of the characteristics of a cube but has octagonal rather than square faces and softer corners (Figure 13.8). This is the **truncated cube.** It fits snugly into a cube, filling 97% of its space.

Like the truncated tetra, the truncated cube has a volume in a seven-part proportion: six parts in octagonal pyramids and one part in triangular pyramids. The plan view of a truncated cube is surprisingly spherical in appearance with an equator in the shape of a dodecagon.

At the center of this series is the **cubocta,** a shape that results from the successive truncation of the square faces of a cube or the succesive truncation of the triangular faces of an octahedron.

The cubocta is the only one of the classical solids with an edge length equal to the radius of its enclosing sphere. The internal partitions of the cubocta are equilateral triangles; the pyramids of the cubocta are tetrahedrons (40% by volume) and half-octahedrons (60% by volume). Both of these are minimum surface pyramids.

The cubocta has two kinds of faces but only one dihedral angle. It is as easy to construct as solids with only one kind of face. Also, the cubocta divides cleanly at an equator, which means that as a half-shape or hemisphere it is complete in terms of its pyramidal structure. Only the octahedron and the icosadodecahedron can be similarly divided.

The cubocta is the shape Buckminster Fuller used for his patented world map (Figure 13.9). The vertices of the cubocta are the centers of closely packed spheres.

A cubocta fits snugly into a cube, filling 83% of the space. The square faces of the cubocta are turned at 45° with respect to the cube, however.

The cube, truncated cube, and cubocta have corresponding two-dimensional patterns of squares and octagons (Figure 13.10). The triangular faces of the truncated cube and the triangular faces of the cubocta correspond with the black squares in the flat patterns.

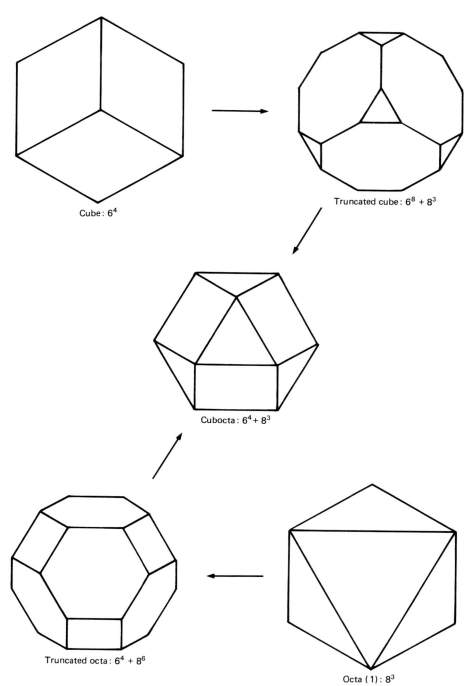

Figure 13.6
The cube-octa series of solid shapes.

Cube: 6^4

Truncated cube: $6^8 + 8^3$

Cubocta: $6^4 + 8^3$

Truncated octa: $6^4 + 8^6$

Octa (1): 8^3

The **octahedron** (Figure 13.11) is a regular solid, a regular antiprism, and a pair of square pyramids. It can be considered a shape in its own right or as the end result of the successive truncation of the triangular faces of a tetrahedron. The

shape is called an octahedron because it has eight faces and not because it has any connection with an octagon.

Resting on a face, an octa is a triangular regular antiprism with a hexagonal equator. Balanced on a

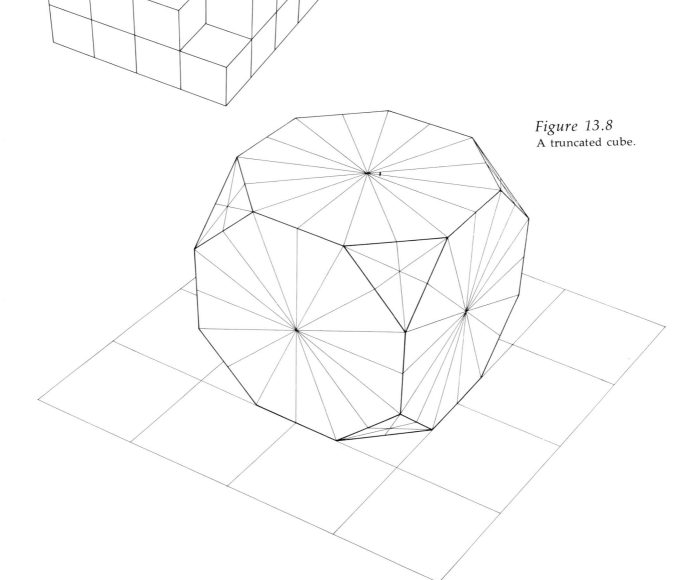

Figure 13.7
The cube. Cubes, of course, can be subdivided into smaller cubes—8, 27, 64, 125, and so on—and any number of cubes can be assembled into larger rectangular volumes. But this process seldom is encountered in buildings.

Figure 13.8
A truncated cube.

Jan. 29, 1946.   R. B. FULLER   **2,393,676**

CARTOGRAPHY

Filed Feb. 25, 1944

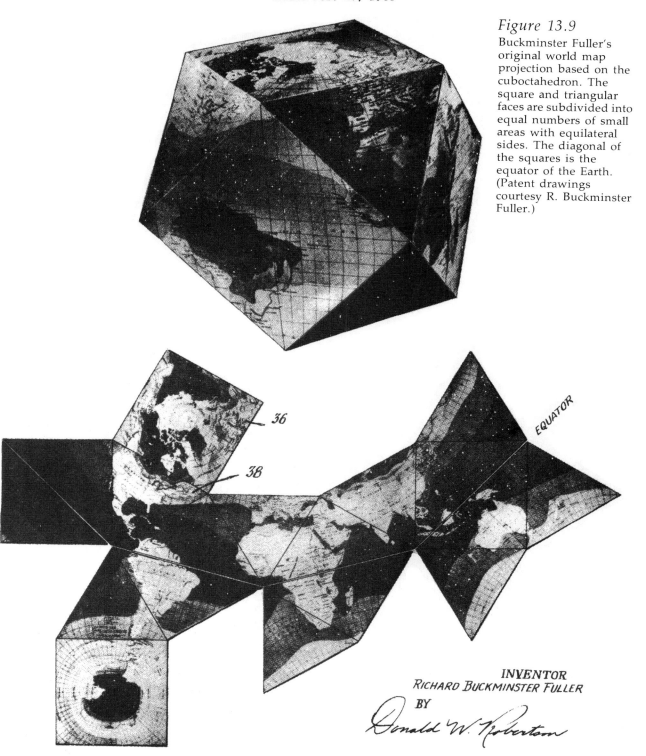

Figure 13.9
Buckminster Fuller's
original world map
projection based on the
cuboctahedron. The
square and triangular
faces are subdivided into
equal numbers of small
areas with equilateral
sides. The diagonal of
the squares is the
equator of the Earth.
(Patent drawings
courtesy R. Buckminster
Fuller.)

36

38

EQUATOR

INVENTOR
RICHARD BUCKMINSTER FULLER
BY
Donald W. Robertson

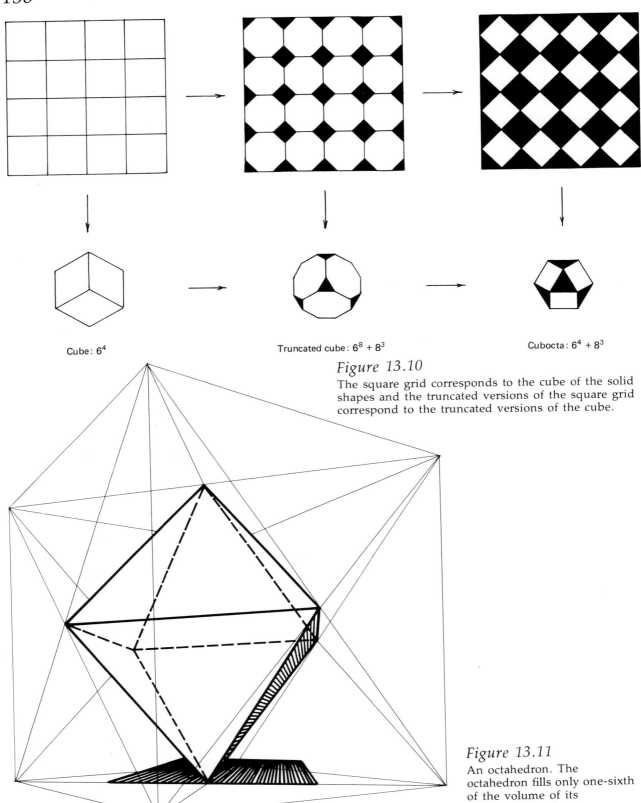

Cube: 6^4

Truncated cube: $6^8 + 8^3$

Cubocta: $6^4 + 8^3$

Figure 13.10
The square grid corresponds to the cube of the solid shapes and the truncated versions of the square grid correspond to the truncated versions of the cube.

Figure 13.11
An octahedron. The octahedron fills only one-sixth of the volume of its surrounding cube.

point, an octa has a square equator. The portion above the equator or the mirror image below is a square pyramid with a ridge slope of 45°, one of the five perfect pyramids. Of the square pyramids, a half-octa has the least exposed surface and because of this might have been the shape chosen by the ancient Egyptian architects for the great pyramids. Not so. The walls of the pyramids of Giza slope at 51° 6' while the slope of the half-octa is 54° 44'.

The octa, cubocta, and others of the solids can be constructed in a variety of ways. Figure 13.12 shows outlines of the shapes held in tension by curved bands of equilateral triangles.

Naturally enough, cubes are thought of as the epitome of the Cartesian plane coordinate system. However, the internal pyramidal structure of the cube does not divide space into Cartesian planes. Only the octahedron, the dual of the cube, has an internal structure with the mutually perpendicular planes of the Cartesian system. The internal partitioning of the octahedron divides space into eight pyramidal blocks with right-angular walls.

There is another point that is interesting about this shape. The surface of an octa represents points at an equal distance from the center when the path of travel is restricted to a three-dimensional rectangular grid. The octahedron is the three-dimensional equivalent of the square turned at 45° on a street grid described in Chapter eight. Points on the surface of a sphere are at an equal distance from the center only if travel is on radial lines.

As the dual of the cube, the octa fits perfectly into a cube but only the vertices touch the faces, a delicate suspension. It fills one-sixth of the space of a cube.

Truncation of the triangular faces of an octahedron produces the eight hexagonal faces of the **truncated octahedron** (Figure 13.13). The remaining spaces are squares.

By volume, 75% of the truncated octa is in hexagonal pyramids and 25% is in square pyramids. Both of the pyramids of the truncated octa are minimum-surface pyramids, the hexagonal when total surface is counted and the square when exposed surface is counted. This same combination of minimum-surface pyramids occurs in the cubocta.

The pyramids of the truncated octa comprise the tepee village shown in Chapter 12. The truncated octa has one outstanding characteristic. Any number of the same size can be assembled to-

Figure 13.12
Models of an octahedron and a cuboctahedron. The outlines of the solids are held in tension by the curved bands of the equilateral triangles. (Photo by Nisham.)

gether like cubes with no leftover space. All of the faces will match. Among the semiregular solids, the shape is unique in this respect. Because the truncated octa is also a relatively good approximation to the sphere (91%), this unique packing feature provides a three-dimensional method of assembling round or near-round objects in a space

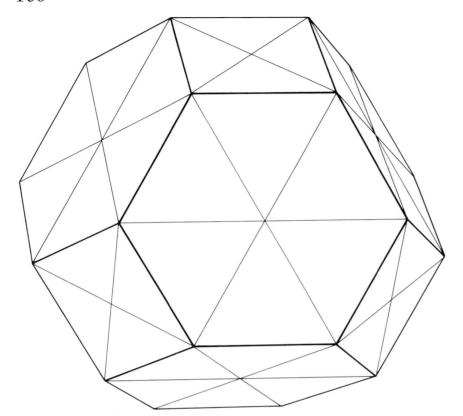

Figure 13.13
A truncated octahedron. The shape is the only solid other than the cube which packs tightly together.

frame made of straight-line elements (Figure 13.14).

Like other solids in the cube-octa series, the truncated octahedron fits snugly into a cube. It fills exactly half the volume of a cube. From one view the equator is a hexagon, and from another, it is an octagon. In either case, the shape does not divide cleanly at an equator in terms of its pyramidal structure.

The truncated octa is also known as Kelvin's polyhedron after Lord William Kelvin, an Irish physicist (1824–1907) noted for his work in pure science.

ICOSA-DODECA SERIES

Five solids with surfaces of pentagons and decagons as well as equilaterals and hexagons comprise another series of shapes linked by truncation, the icosa-dodeca series (Figure 13.15).

There are no squares in this series and none of the solids has any relationship to the cube of the

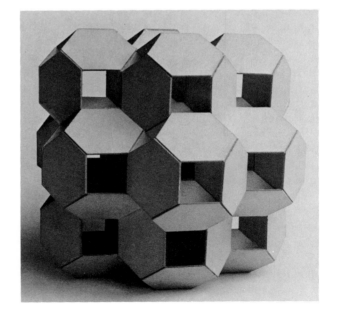

Figure 13.14
An assembly of truncated octahedrons. (From *Shapes, Spaces and Symmetry* by Alan Holden, Columbia University Press. Photo by Doug Kendall.)

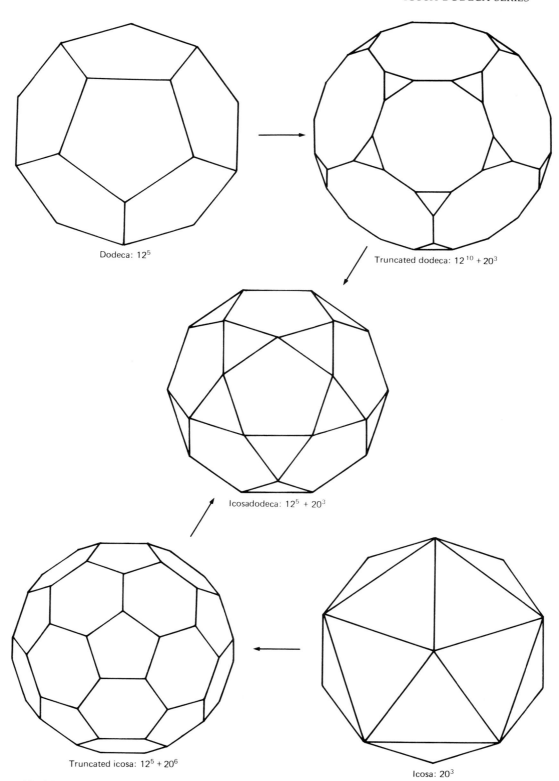

Dodeca: 12^5

Truncated dodeca: $12^{10} + 20^3$

Icosadodeca: $12^5 + 20^3$

Truncated icosa: $12^5 + 20^6$

Icosa: 20^3

Figure 13.15

The icosa-dodeca series of the classical solids. These five shapes are also linked
by the process of truncation.

Cartesian coordinate system. The dodeca and the icosa establish a plane system of their own.

The five solids are all close approximations of the sphere, so close that one of them, the truncated icosahedron, is the shape of a soccer ball.

The series begins with the **dodecahedron,** a shape with twelve pentagonal faces and a ten-sided equator, a decagon. The twelve pentagonal faces of the dodeca can carry the signs of the zodiac as well as the months in a year. The shape of the dodeca places these signs in their most influential and compelling relationship and is perhaps the only true representation of the astrological forces. On a different plane, this shape is the point of departure for an intricate sculpture, *Eclipse,* by Charles Perry (Figure 13.1), in which the interpenetration of space conveys the sense of a whole universe geometrically defined.

As may be seen (Figure 13.16), the dodecahedron has a strong affinity with the flat hexagonal tile pattern, its two-dimensional counterpart.

The **truncated dodecahedron,** in which the pentagons of the dodeca have become decagons, offers the greatest contrast in size between two shapes on the surface, tiny equilaterals pinning spacious decagons. The decagons can be subdivided into "sublime" triangles to generate linework on the surface of increasing complexity and interest.

If truncation continues, the next shape in the series is the **icosadodecahedron.** The icosadodeca combines the twenty triangular faces of the icosahedron with the twelve pentagonal faces of the dodecahedron and can be obtained by the successive trunction of either of these shapes. It occupies the same position in the icosa-dodeca

Figure 13.16
A dodecahedron shown with the hexagonal floor grid, its two-dimensional equivalent. (Photo by Gerald Ratto.)

series as the cubocta occupies in the cube-octa series. Like the cubocta, the icosadodeca has only one dihedral angle.

The icosadodeca has circumferential lines corresponding exactly to the great circles of the sphere. In terms of its interior pyramidal construction, it divides cleanly at the equator, the only shape in this series to do so. A half-icosadodeca is a 95% approximation to a true hemisphere and much easier to construct. Figure 13.17 shows the half-octa, half-cubocta, and half-icosadodeca, the three solids that can be naturally halved at the equator. A hemisphere with the same volume is included for comparison.

The pentagonal pyramids of the icosa-dodeca are the folded-up points of the star (Chapter 5); the partitions of this solid are "sublime" triangles. In one way or another, all five shapes in this series embody the number of the golden mean, ϕ.

Twenty equilaterals make up the **icosahedron,** another of the solids with a ten-sided equator. The icosa is the spherical equivalent of the 60° grid and all of the many patterns of the 60° grid can be incorporated on its surface. The equilateral faces, of course, can be subdivided into modular equilaterals of any size. The midsection of an icosa is a regular pentagonal antiprism and the caps are pentagonal pyramids. The "icosa" pavilion of Chapter 5 is an icosa with a ring of smaller pentagonal pyramids (icosacaps) at the equator.

The **truncated icosahedron,** the shape of a soccer ball, combines pentagons and hexagons on the surface. The vertices are relatively flat rather than pointed, which gives the shape its rounded appearance. The hexagons are the truncated equilaterals of the icosahedron. If truncation continues in

Figure 13.17
The half-octa, half-cubocta, half-icosadodeca, and hemisphere. All of the shapes have the same volume. (Photo by Gerald Ratto.)

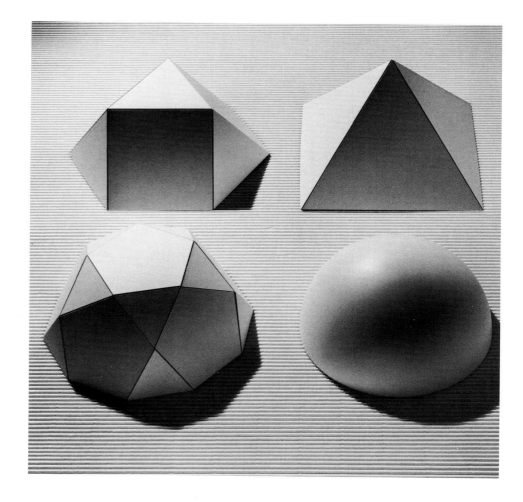

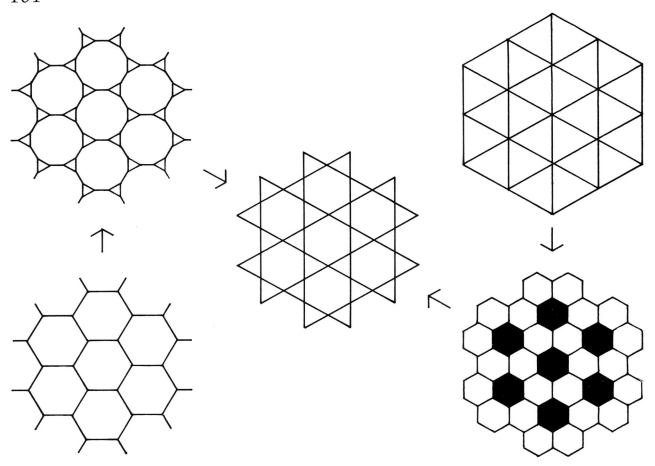

Figure 13.18
Hexagonal and equilateral triangular tile patterns linked by the process of truncation.

the same direction from the same points, the end result will be the icosadodeca at the center of Figure 13.15.

The five 60° tile patterns linked by the process of truncation shown in Figure 13.18 correspond exactly to the five solids in the icosa-dodeca series.

The only difference between the geometric solid and the corresponding flat pattern is that the pentagons of the solid are hexagons in the flat and the decagons of the solid are dodecagons in the flat. This is a very slight difference between round and flat, between a three-dimensional object and its two-dimensional counterpart.

The process of truncation in the tile patterns is identical to the process of truncation in the corresponding solids.

A tile pattern, of course, can be extended indefinitely but the solid shape is limited to a finite number of pieces: twelve, twenty, or thirty-two in the icosa-dodeca series.

THE MASTER CHART

The twelve geometric solids and the five patterns described to this point are not only linked by the process of truncation; they also fall into a number sequence of 3–4–5 and 6. The master chart (Figure 13.19) shows these plane and solid shapes drawn to scale in their proper sequential relationship, both horizontally and vertically.

The top view of the solids is shown. The shapes and patterns in Columns A and E are the regular solids and regular plane figures. Columns B, C, and D are truncated versions of these shapes.

Column E includes constructions of equilateral triangles. These triangles come together at a point

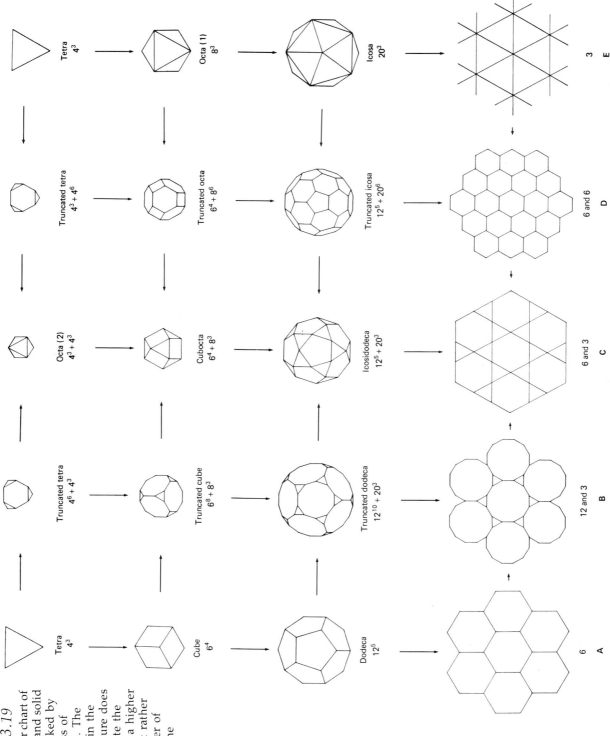

Figure 13.19
The master chart of the plane and solid figures linked by the process of truncation. The exponent in the nomenclature does not indicate the raising to a higher power but rather the number of sides in the polygon.

in a sequence of 3–4–5 and 6, corresponding to four of the five "perfect" pyramids. In Column D, the equilaterals are truncated to hexagons but the number of them doesn't change.

The 3–4–5–6 sequence occurs again in Column A, but this time it is the number of sides in the polygons of each shape, not the number of equilaterals meeting at a point. The truncated versions in Column B have twice the number of sides and a number sequence of 6–8–10–12 rather than 3–4–5–6.

Column C, the center column, combines features of both the left and the right sides. The solid shapes in this column are the only ones that divide cleanly at the equator in terms of the internal pyramidal construction.

The visual relationships between the shapes on the master chart are plain to see. Note that the top row reads the same from right to left or from left to right.

THE SIX REMAINING CLASSICAL SOLIDS

There are six geometric solids that are not linked by the process of truncation. They are the great and small rhombicubo, the great and small rhombicosi, the snub cube, and the snub dodecahedron. The faces of these solids are regular polygons of three, four, five, six, eight, and ten sides.

All six of the remaining solids are very close approximations to a sphere. The **snub dodeca** (Figure 13.20), the only one without a square face,

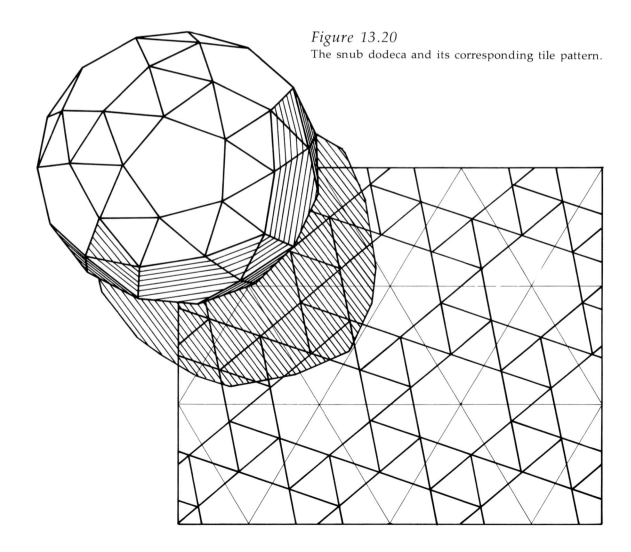

Figure 13.20
The snub dodeca and its corresponding tile pattern.

is the most efficient volume enclosure of all the classical solids, a 98.2% approximation to a sphere. Although it has the least surface area required to enclose a given volume, it has the most faces (92). A snub dodeca has associated with it a flat pattern of hexagons and equilaterals with an identical pattern of assembly. This pattern is shown with the snub dodeca in Figure 13.20.

The **snub cube** (Figure 13.21) is a formless arrangement of squares surrounded by equilaterals. It is a small shape with little resemblance to a cube and is without a corresponding flat pattern. If there is an unwanted child in the family of geometric solids, it is the snub cube.

The **small rhombicubo** (Figure 13.22), on the other hand, is powerfully structured, easy to construct, and retains an obvious relationship with a cube and the Cartesian coordinate system. In plan and in section, it is an octogan and in this respect is similar to the constructions approximating a sphere in the progression of fourths.

The **great rhombicubo** (Figure 13.23), the first of the solids with faces of three different polygons, presents a progression of polygons with an even

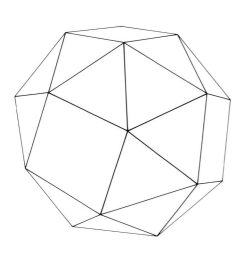

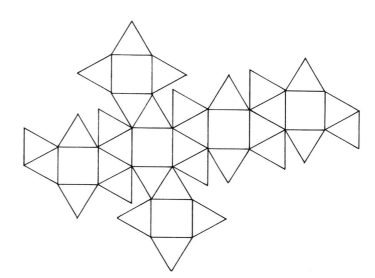

Figure 13.21
The snub cube and its pattern of construction.

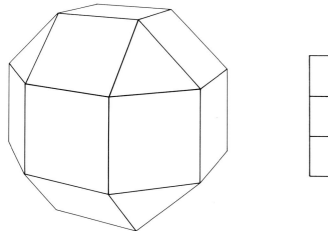

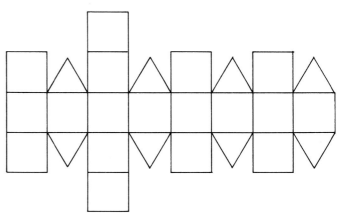

Figure 13.22
The small rhombicubo and its pattern of construction.

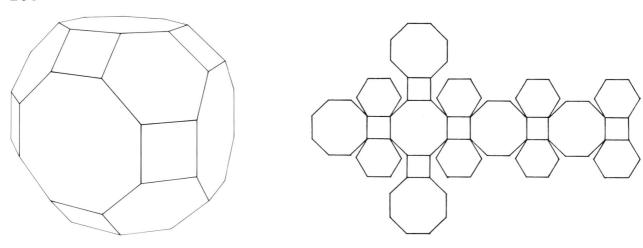

Figure 13.23
The great rhombicubo and its pattern of construction. The tile pattern associated
with the truncated cube (Figure 13.10) is similar but rotated 45°.

number of sides, four, six, and eight. The outstanding architectural attribute to this shape is
that it retains unmistakable cubelike qualities in
spite of the number of different sides. The great
rhombicubo is a transformation of the cube and it
fits perfectly into a cube, filling 74.5% of the space.
Because of the dominance of the six octagonal
faces, the great rhombic establishes an unequiv-
ocal parallel relationship with Cartesian planes
when it is suspended in space. When the shape
is hung in the center of a room, the octagonal
faces generate a pulsating rapport with the walls,
floor, and ceiling of the room.

The great rhombicubo has a cumbersome name
(rhomicuboctahedron) and appears complicated,
but the formulas for its surface, volume, and
pyramidal construction reduce to simple, clear,
and elegant expressions in terms of the square root
of two.

It can be compared directly with the larger and
grander **great rhombicosi,** which holds the same
position in the 5/10 complex of shapes as the great
rhombicubo has in the 4/8. Both shapes have similar compositions but the rhombicosi has many
more faces, sixty-two compared to twenty-six, and
uses a decagon instead of an octagon as its principal shape. While the volume of the great rhombicubo and all of its interior pyramids can be simply expressed in terms of the square root of two,
the volume of the great rhombicosi and its
pyramids can be simply expressed in terms of the
square root of five. The great rhombicosi relates to

the twelve planes of the dodeca rather than the six
planes of the cube.

The great rhombicosi (Figure 13.24) is a magnificent shape without equal in the realm of assemblies of the regular polygons. It has a floor
pattern to match. A great rhombicosi has more
vertices and edges than any of the other solids and
it has the steepest interior partitions.

The **small rhombicosi** (Figure 13.25) is the only
shape that combines the primary polygons—
equilateral, square, and pentagon—into one surface. Like the great rhombicosi, it, too, has a flat
pattern associated with it.

THE POLYGONS OF THE SIX
REMAINING SOLIDS

The regular polygons of the six remaining solids
are unrelated by truncation but they must all have
the same edge length if they are to fit together
(Figure 13.26). This sequence of six polygons
excluding only the seven- and nine-sided polygons (which seldom appear anywhere in architectural geometry let alone in the classical solids)
illustrates an important relationship between a
regular polygon and its double.

Observe that the point of the equilateral triangle
is at the exact center of the hexagon and the point
of the pentagon is at the exact center of the decagon. If the square were rotated 45°, it, too, would

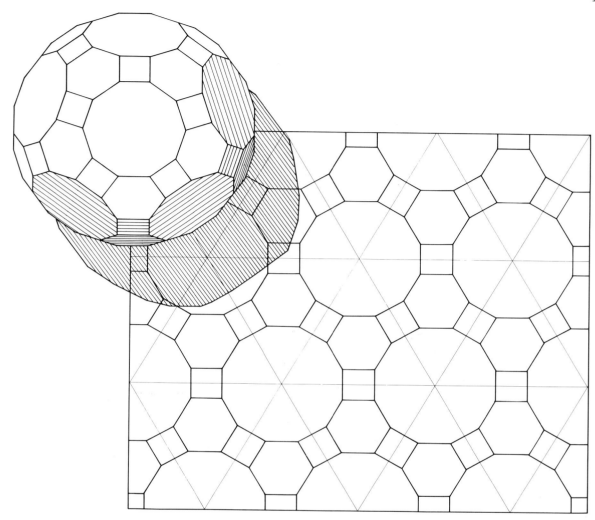

Figure 13.24
The great rhombicosi and its corresponding flat tile
pattern. The decagons in the solid shape are
dodecagons in the flat pattern.

have a point at the exact center of its double, the
octagon.

In trigonometric terms, the inside radius plus
the outside radius of one regular polygon will always equal the inside radius of a second polygon
with twice the number of sides. The length of side
is the same in both polygons.

This relationship leads to two (or more) endless
series of increasingly intricate geometric constructions of unusual symmetry and beauty (Figure
13.27). The first places the smaller of the two polygons in each possible position on the inside
perimeter of the larger. When the smaller polygon

has an even number of sides, there will be an eye
at the center of the figure. The second series centers the larger polygon at each possible position
around the smaller.

Designs involving line tracery such as these offer
endless fascination to the human eye and mind
and, in that way, become visually dynamic, rather
than static, figures. When these geometric constructions are used as frames to hold pieces of carefully chosen colored glass to the light, another dimension, the endless variety of daylight and cloud
cover, is added. The visual effect as seen in the
windows of cathedrals can be wondrous.

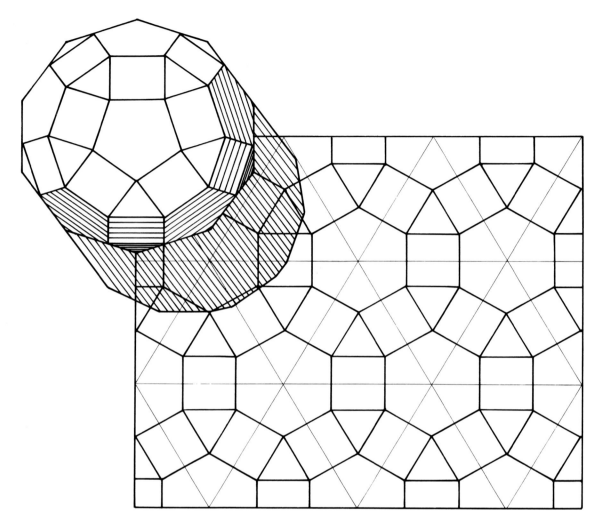

Figure 13.25
The small rhombicosi and its corresponding tile pattern. The pentagons in the solid are hexagons in the flat pattern.

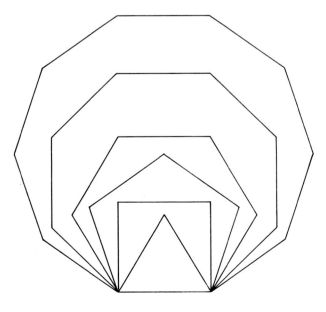

Figure 13.26
The six regular polygons which make up the faces of the six remaining classical solids. To be interchangeable, they all must have the same edge length as shown in this drawing.

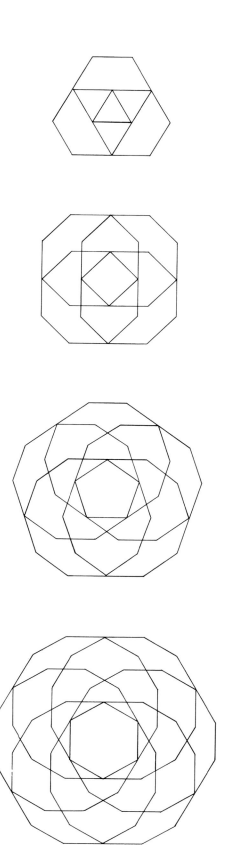

Figure 13.27
Two series of
designs that follow
from a polygon and
its double, that is,
one with the same
edge length but
twice the number
of sides.

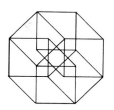

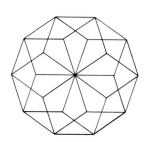

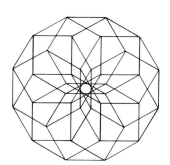

CHAPTER 14

DOMES, SPHERES, AND HEMISPHERES

SPHERE: The apparent surface of the heavens of which half forms the dome of the visible sky.

Webster's Third New International Dictionary

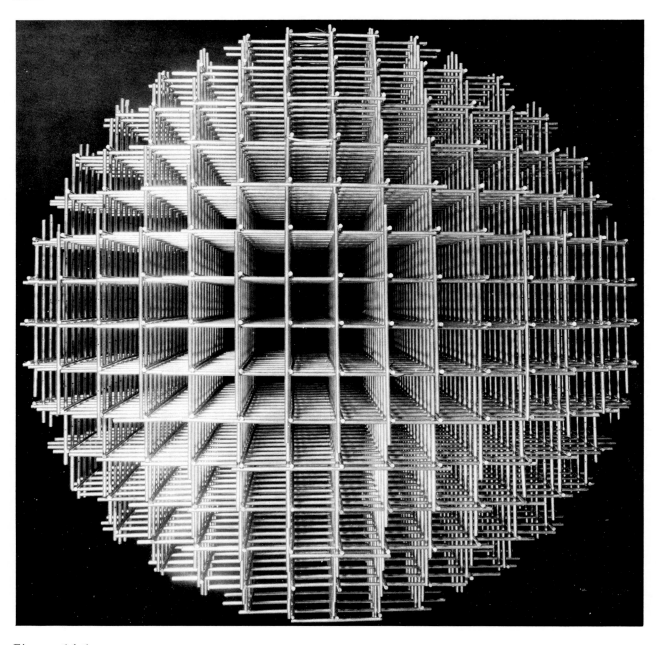

Figure 14.1
Sphere-Web, a lattice sphere of metal rods by the sculptor François Morellet, 1967.
(Courtesy Galerie Denise Rene, Paris.)

THE CULMINATION OF THE CLASSICAL SOLIDS is a sphere. It is the one shape with every point on the surface equidistant from the center. It is the most compact of all of the solids, and it is the most efficient of the solids in terms of the amount of material required to enclose a given volume. In this respect, it is the measure of all other shapes.

Seldom, however, can an architect take advantage of the surface efficiency of the sphere. While hemispheres and domes are seen frequently, a building in the shape of a complete sphere is rare. There are several reasons for this, all of them fundamental to architecture.

In the first instance, a horizontal surface (floor) is the first essential of a completed building, and the floor of a sphere is not horizontal. In fact, the slope of the surface of a sphere is everywhere changing, vertical at the equator, flat at the poles, and every other slope inbetween.

Because of the constantly changing slope of the walls, a true sphere is difficult to construct of linear materials. Much ingenuity has been spent in developing solutions to this problem. This is apparent in the many methods of constructing hemispheres and domes available to the architect and engineer. Figure 14.1 is a unique rectilinear approximation to a sphere.

Moreover, the sphere is the shape of a ball which can be rolled on the lawn, but this feature is of little value to buildings intended to be stationary. The Ledoux House (Figure 14.2) looks as though it could, indeed, roll away. At the least, the image conveys a sense of impermanence in position.

The sphere is a superb container for gases which exert equal pressures in all directions and a superb container to withstand such pressures from the outside, but these pressures are not normally encountered in constructions intended for human occupancy on the planet Earth.

The geometry of the whole sphere is very appealing but, because of the limitations of architecture, satisfaction must be found in the half-sphere or hemisphere, the dome of the visible sky, not the whole surface of the heavens.

HEMISPHERES AND DOMES

Hemispheres and domes are used to enclose large column-free areas because of the structural strength of the curved surface. Originally they were made of masonry (or ice, in the case of igloos) and like a semicircular arch were capable of supporting heavy weight across a short distance. An array of domes covers the Mosque of Ahmed I (Figure 14.3).

The shape of the area enclosed by a hemisphere is, of course, a circle. If the building is round, as is the Pantheon in Rome (Figure 14.4), support for the dome is provided directly by the walls under

Figure 14.2
A house in the shape of a sphere by Claude-Nicolas Ledoux, after 1773. (Courtesy of the Art & Architecture Library, Yale University.)

Figure 14.3
The roof domes of the
Mosque of Ahmed I.
(Photo by G. E. Kidder
Smith.)

Figure 14.4
The Interior of the Pantheon, painting by Giovanni Paolo
Panini, c. 1750. (From the Samuel H. Kress Collection,
National Gallery of Art, Washington, D.C.)

it, but when the floor area is rectangular, a structural transition must be made from a circle to a square. Historically, this transition has been made by great spherical pendentives geometrically based on the cuboctahedron, as at Hagia Sophia (Figure 14.5).

The problem of placing a round dome on a square base can be solved in other ways, however. Figures 14.6 and 14.7 show a different solution. In this concept, emphasis is placed on geometric transitions in stages, first from the circle of the dome to a larger circle, thence to a cylinder which is cut in 45° elliptical planes to make the transition to a square. As in Hagia Sophia, the width of the square base is equal to the diameter of the dome.

The model illustrates transition elements, which are as important in architecture as they are in music or any of the arts. The way in which a change is made from one shape to another is a vital part of the geometry of architecture.

CONTEMPORARY DOMES

In contrast to ancient masonry domes, those built today for stadiums and other purposes are of

Figure 14.5
The interior of Hagia Sophia, Istanbul. (Photo by Hirmer Fotoarchiv, Munich.)

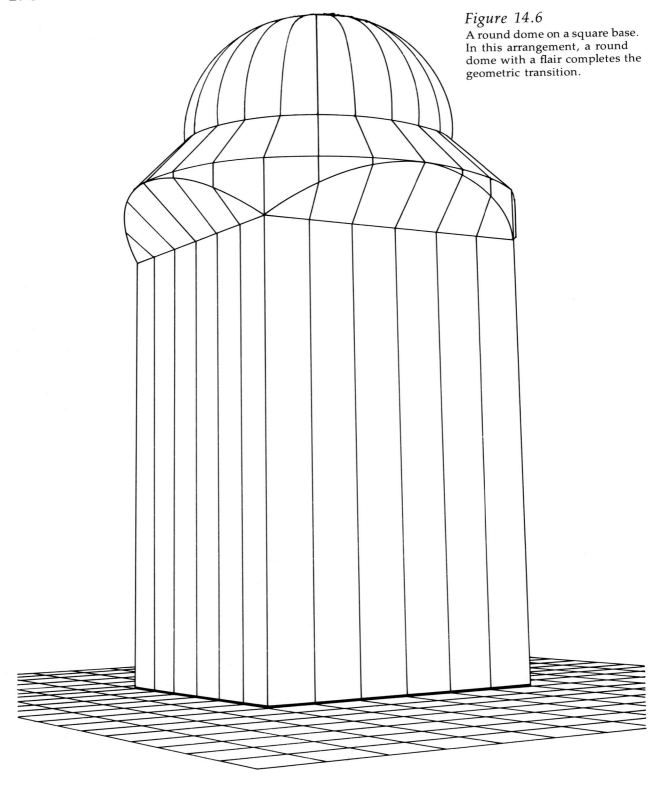

Figure 14.6
A round dome on a square base.
In this arrangement, a round
dome with a flair completes the
geometric transition.

lightweight materials and cover much larger areas. They are placed directly on the ground and are not just a part of a building; they are the whole building (Figure 14.8). Consequently, the problem of a structural transition to a rectangular framework disappears. The limits of size have not yet been finally established. Buckminster Fuller, who single-handedly sparked renewed interest in the shape, once proposed a hemispherical dome two miles in diameter to enclose a part of New York City. His object was to provide economical environmental control including air conditioning for a large part of the city. This concept has not been tested, but Buckminster Fuller's development of the geodesic dome has contributed greatly to the excitement and interest of contemporary architecture.

As roof shapes, hemispheres provide a unique problem because the slope is constantly changing from vertical at the equator to horizontal at the apogee. Conventional roofing materials are designed for one range of roof slopes, not for every slope. Materials covering a hemisphere must be adaptable to the full range of slopes, or else the roof of the dome must be divided into bands or zones, each one suitable for a particular material. The structural zones of one kind of lightweight, three-dimensional roof structure are seen in Figure 14.9.

As space enclosures with equal volume, it is interesting to compare a hemisphere with a half-octahedron, the most effecient of the square pyramids. Between the two, the hemisphere has less exposed surface, is lower in height, and has a soft, rounded shape with a constantly changing slope, is circular in plan at every level, and can be constructed with interlocking pieces as in the geodesic domes or by the method of successive approximations with panels of uniform width. The pyramid is higher, has a constant angular wall slope, more exposed surface, square floor plans at every level, and a modular triangular assembly system. Both the hemisphere and the pyramid can provide huge column-free interior spaces.

A part of the attraction of the hemisphere is its shape. The dome of the Pantheon is said to resemble the curved canopy of heaven. The world is round. As far as the eye can see, the universe is

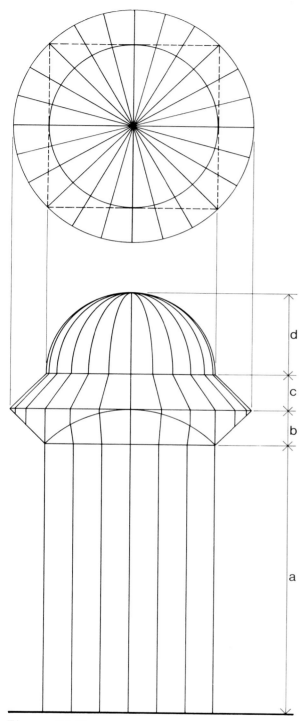

Figure 14.7

The top and side view of a dome on a square base. The linework is continuous and traceable from top to bottom.

Figure 14.8
The geodesic dome of R. Buckminster Fuller enclosing the U.S. Pavilion, Expo '67
World's Fair, Montreal. (Photo by Joseph W. Molitor.)

round. The similar shape of the hemisphere conveys wordly and universal concepts, which is why it is often used for academic institutions.

A hemisphere obviously is a perfect shape for a planetarium. Every point on the surface is equidistant from the projectors at the center. Conversely, the round walls of the hemisphere direct rays of light, sound, and vision to the geometric center, placing extraordinary importance on that point when other kinds of uses are considered.

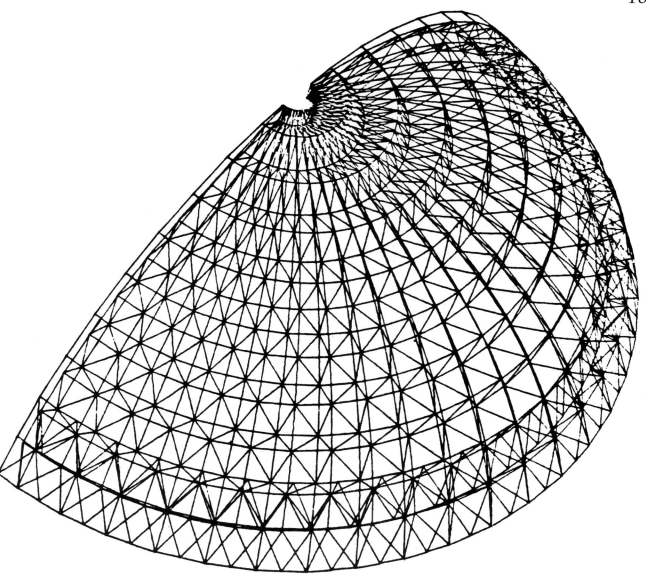

Figure 14.9
A computer drawing of the structure of a dome. Half only is shown. (Courtesy
Leon Nadalski and Istvan Kadar, Bechtel Civil & Minerals, Inc.)

Surprisingly, domes seldom are used for shopping centers, the marketplace of today, although the central space of a shopping center, analogous to a city square and with much the same geometry, presents an opportunity for a special architectural solution such as the dome.

A complete hemisphere may provide too much height in relation to the area enclosed, and is limited to a circular shape in plan. Partial domes are lower and can cover triangular, square, or rectangular areas as well as circular areas. Often they are more economical than full hemispheres. The Kresge Auditorium at M.I.T. (Figure 14.10) is an example of a partial spherical surface enclosing a noncircular interior space. Other relatively flat domes, including air-supported fabric structures, cover huge athletic stadiums of today.

Figure 14.10
The Kresge Auditorium at M.I.T. by Eero Saarinen, 1955. (Photo by Joseph W. Molitor.)

CONCLUSION

This chapter on domes, spheres, and hemispheres completes the review of the basic shapes of architecture. The review began with the regular polygons and progressed through the right triangles, rectangles, ellipses, prisms, pyramids, and other shapes encountered in the design and construction of our environment. It presented the geometries by which things are made.

As on any journey, only a limited number of events can be seen, absorbed, and recorded. The recurrence of certain numbers such as ϕ, the progression of fourths, the simple but effective diamond and diagonal geometry of the streets, the importance of perimeter, the surprising relation-ship between wall and floor area for maximum volume enclosure, and the rediscovery of proportion have been some of the high points.

Surely, some important things have been missed but just as surely someone will find them and add new chapters and new books. The process of discovery and formulation will continue. Indeed, the subject of geometry in architecture may be inexhaustible, its vitality a constant in the life and works of human beings. Hence, while the book is at an end, the subject is not.

The rewards of a journey are in the experiencing of it and in the sharing of it. In this instance, the rewards have been ample.

I N D E X